96 Advances in Polymer Science

Filled Polymers I

Science and Technology

Editor: N. S. Enikolopyan

With contributions by
N. S. Enikolopyan, M. L. Fridman,
A. Y. Malkin, N. E. Nikolaeva,
A. T. Ponomarenko, V. L. Popov,
O. Y. Sabsai, V. G. Shevchenko,
I. O. Stalnova

With 34 Figures and 21 Tables

Springer-Verlag
Berlin Heidelberg New York London
Paris Tokyo Hong Kong Barcelona

Guest-Editor:

Prof. N. S. Enikolopyan
USSR Academy of Sciences
Institute of Synthetical
Polymeric Materials
Ul. Profsojuznaja 70
USSR-117393 Moscow

ISBN-3-540-52791-5 Springer-Verlag Berlin Heidelberg New York
ISBN-0-387-52791-5 Springer-Verlag New York Berlin Heidelberg

Library of Congress Catalog Card Number 61-642

Typesetting: Th. Müntzer, Bad Langensalza; Printing: Heenemann, Berlin; Bookbinding: Lüderitz & Bauer, Berlin
2152/3020-543210 – Printed on acid-free paper

Editor's Foreword

The development, the study of their properties and putting highly filled polymer composites to commercial use on a large scale in various branches of industry has, for the last 10 or 15 years, become the "strategy" of the chemical technology of polymers and their processing into end products. This, above all, holds good for industrial thermosoftening plastics filled with dispersed and fibrous substances of mineral, organic and mineral-organic nature.

Production of highly filled polymer composites (HFPCs) and their processing into all kinds of end products have long been known and are widely used in industry. As an example, composites of all descriptions based on thermosetting resins or HFPCs produced on the basis of plasticized PVC. Thus, linoleum produced by extrusion is made of PVC, containing 25 to 45 parts by weight of a low-molecular plasticizer and 60 to 110 parts by weight of a finely-dispersed mineral filling agent (usually, calcium carbonate) per 100 parts by weight of a binding agent (matrix). The term "ballast" which used to denote the fillers has grown obsolete as it has become common to regard a filler, in most cases, as capable of being "active", viz. of changing the set of properties of plastics. In particular, this applies to the ability of fillers to enhance the plastics' strength characteristics, chemical resistance, rigidity, heat resistance, to reduce shrinkage, thermal expansion, to broaden the range of permissible operating temperatures, etc. It is common knowledge that the required quality parameters of most items made of thermosetting plastics can only be achieved by adding to resins the fillers that should be viewed as modifiers of the properties of materials rather than anything else. This applies equally to thermoplastic: selective control of their properties until recently was thought to be the major object of filler application, whereas the resultant saving of polymer material ("base" polymer) was regarded as nothing but a natural side-effect. In cases where the saving of a polymer material was the main objective, and a large quantity of a filler was placed in the matrix, the degradation of mechanical properties of materials was accepted as an inevitable "evil" one has to put up with. All this applies to thermosoftening plastics since the above two objectives had long been incompatible. Experts had sought (and frequently succeeded in finding) compromise recommendation such as "Do not add more filler than this certain quantity", for manufacturers of composites. They also suggested that end-product users try to economize on the polymer by trying to get agreement on the relaxation of requirements for strength and durability.

Of late (roughly, since 1973 or 1974), the problem of the polymer filler has again become topical. To begin with, there has been mounting interest in the use of fillers with a view to reducing polymer requirements and bringing down the cost of thermosoftening plastics. The latest achievements in the fields of physics-chemistry, mechanics, rheology and polymer processing technology have laid the foundation for resolving new fundamental problems in the area of filling the thermoplastics, namely: securing a considerable saving of base polymers without impairing the set of major performance characteristics of materials; development of highly filled composite systems that feature deformation-strength parameters and durability similar to or better than those of pure polymers; development of highly filled thermoplastics that can be used as structural materials.

It is anticipated that in future, economic advantages of HFPCs will be put into reality through priority growth of filler production compared with polymer production. The following technico-economic factors will be conducive to this. First, the cost of fillers should increase slower than that of polymers, because production of most filling agents has nothing to do with the depleting oil reserves and, besides, it is far less energy-intensive than production of polymer materials. During the four-year period alone (from 1973 to 1976), world market prices of oil doubled, while the prices of fillers increased by only 25 to 30%. As the gap between these prices grows steadily wider, the economic expediency of using cheap fillers will be increasingly apparent. Second, the number of machines and the number of plants capable of processing various HFPCs into high-quality end products will be expanding. Finally, new uses of end products and semi-finished articles made of composite materials continually "crop up", viz. areas where such uses were previously considered not only unthinkable but impossible. There is hope that the above factors will serve to encourage improvement of the 'psychological atmosphere' for further development of processing and new applications of HFPCs.

The term "filler" is fairly broad to denote quite a variety of substances. Usually, the term means solid substances added to polymers in relatively large quantities, i.e. 5% of mass. Such a definition does not include substances used in low concentrations:

lubricants, dyes, stabilizers, antistatics, liquid filling agents, plasticizers, etc. even though the importance of these ingredients is generally acknowledged. Most of the traditional dispersed fillers discussed in the chapters of this book are minerals extracted from rocks or ores and transformed to powder after being subjected to appropriate treatment.

A substantial modification of practically all the properties of polymers is achieved by adding 10 (or more) volumetric % of dispersed filling agents, and for this reason composite materials like this may generally speaking, be grouped with highly filled polymer composites (HFPCs). Such gradation is made for other reasons, too. The point is that production of HFPCs necessitates a number of additional operations (preparation of fillers, mixing of components, etc.); on the other hand, HFPCs processing into end products is made difficult by high viscosity and the fast wear of equipment by filled composites as compared with non-filled systems. Therefore, economists estimate that the use of dispersed fillers is only profitable when a filler is added at the rate of not less than 10 to 15% by volume.

This book includes some papers which, in our view, contribute to the laying of a scientific foundation for the solution of the tasks formulated above.

The existing scientific basis, the massive "assault" being launched by researchers and engineers in many countries at present as well as available experience gained in the earlier attempts to find ways of resolving the above-named problems enable us to share the optimism of many experts (e.g. see "Fillers of Polymer Composites" ed. by G. S. Katz and D. V. Milevsky and the encyclopaedic series "Composite Materials" in 8 volumes, ed. by L. Broutman and R. Krok published by Academic Press). According to the experts, the time is not too far away when production of articles from non-filled polymers will, more often than not, be less expedient technically and economically than manufacture of same articles from filled materials, the efficiency of HFPC utilization being constantly on the upgrade.

The book contains rather complete reviews of papers published in the last 5–10 years in the USSR and abroad on various problems of filled polymers of differing nature. The discussion is centered on the physico-chemical problems of these complex systems, their structure, mechanical, rheological, dielectric and other properties: in a word, important aspects of theory and technology of filled composites. We hope the topical nature of the subjects discussed and the selection of authors that appear in the book all help to throw more light on this area of science and technology. The interested reader will be able not only to appreciate the book as a source of additional literature or a "snapshot" of the state-of-the-art of the problem, but also as an attempt by the authors to "bridge the gap" between scientific fundamentals and development of filled polymer system technology on the one hand, and effective industrial applications of materials representing a huge class in its own right in the various general and special fields.

Moscow, September 1990 N. S. Enikolopyan

Editors

Table of Contents

Filled Polymers: Mechanical Properties and Processability
N. S. Enikolopyan, M. L. Fridman, I. O. Stalnova,
V. L. Popov 1

Rheology of Filled Polymers
A. Y. Malkin 69

Rheology of Gas-Containing Polymer Systems
O. Y. Sabsai, N. E. Nikolaeva, M. L. Fridman 99

Formation Processes and Properties of Conducting Polymer Composites
A. T. Ponomarenko, V. G. Shevchenko, N. S. Enikolopyan 125

Author Index Volumes 1–96 149

Subject Index 165

Filled Polymers: Mechanical Properties and Processability

N. S. Enikolopyan, M. L. Fridman, I. O. Stalnova, V. L. Popov
USSR Academy of Sciences, Institute of Synthetical Polymeric Materials, Ul. Profsojuznaja 70, USSR-117393 Moscow

In this review the wide spectrum of properties of filled polymer systems is generalized. Classical and modern conceptions dealing with the influence of the boundary layer and filler on melting and crystallization of the polymer matrix. Polymer properties in the melted and condensed state are considered, attention being given to the influence of the filler particle size and shape, the nature of its surface, its structure, and agglomeration on the rheological and mechanical characteristics of composites.

Data on structure and properties of filled thermoplastic composites are summarized for the first time. Special attention is paid to the peculiarities of composites with modified polymerized filler behavior during rheological and mechanical testing and processing of such composites.

Introduction 3

1 Effect of a Filler on Polymer Properties 4
1.1 Methods for Estimating the Filler Effect on Polymer Matrices . . . 4
1.2 The Interphase Concept 7
1.3 Crystallization and Melting of Polymers at Interphases 11
1.4 Effect of Interphase on Physico-Mechanical Properties 14
1.5 Effect of Filler Particle Size and Shape on the Physico-Mechanical Properties or Composites 17
1.6 Effect of Filler Particle Size and Shape on Composite Rheology . . 22
1.7 Filler Effect on Resilience of Polymer Melts 27
1.8 Effect of Filler Structurization and Agglomeration on Filler Polymer Properties 29
1.9 Effect of Adhesion on PCM Properties. Property Control by Modification 33

2 Properties of Polymerization-Filled Composites and Composites with Polymerization-Modified Fillers 42
2.1 Preparation of Polymerization-Filled Composites 42
2.2 Structure of Polymerization-Filled Composites 42
2.3 Properties of Polymerization-Filled Composites 46
2.4 Physico-Mechanical Properties of Composites with Polymerization-Modified Fillers 50

Advances in Polymer Science 96

2.5 Rheological Properties of Composites with Polymerization-Modified Fillers 53

3 Some Special Effects in the Flow of Filled Polymers 55

4 Some Recommended Practices in Processing Highly-Filled Compositions 56

5 References 59

Introduction

Polymeric composite materials (PCM) belong to one of the most important classes of structural materials attracting ever growing interest, as manifested by the huge flow of published scientific and technical information dealing with the problems of developing, preparing, processing and using PCM [1].

Whereas the oil and gas fuel crisis has receded recently, the problem of their restrained use still remains an urgent one if only because the oil and gas reserves are not unlimited and practically non-reproducible. Estimates by specialists have shown [2, 3] that the development of PCM with a view to cut the consumption of polymeric materials can only be economically feasible if not less than 15% by volume of the filler are added.

Waste products from a number of commercial processes can be used as cheap and readily available fillers for PCM. For example, lightweight structural materials may be obtained by filling various low-viscous resins with waste materials [4, 5]. Also by adding fillers to reprocessed polymers it is possible to improve their properties considerably and thus return them to service [6]. This method of waste utilization is not only economically feasible but also serves an ecological purpose, since it will help to protect the environment from contamination. The maximum percentage of the filler should in these cases be such as to assure reliable service of the article made from the PCM under specified conditions for a specified period of time.

However, the chief purpose of introduction of fillers into PCM is to make possible the modification of polymers and thereby create materials with a prescribed set of physico-mechanical properties, and, obviously, the properties of filled materials may be controlled by, for example, varying the type of the base polymer (the "matrix") and filler, its particle size distribution and shape. It may not require a large quantity of filler [7]. Thanks to considerable advances in PCM research, their use in a broad range of industries – machine building, construction, aerospace technology, etc. – has become extensive [8–11].

Development of an assortment of polymeric composite materials, decisions concerning their possible scope of applicability and service conditions involve research into a very broad circle of problems:

– choice of the matrix, filler, functional adjuvants and viscosity depressants, finding the optimum PCM formulations to obtain materials with the desired properties;

– determining the possible ways of processing a particular PCM into articles; finding the process parameters that would provide for the best tradeoff between the positive properties of each of the components of the PCM;

– prediction of the serviceability of an article under prescribed conditions and its service lifetime under these conditions.

Solution of these problems is possible without a profound investigation into each and every property of the composite material, estimation of the contribution of each of the PCM components to the product quality, development of an algorithm for purpose-oriented control of product quality by properly adjusting the PCM formula, or employing different physical and/or chemical effects during preparation and machining processes, or optimizing the process conditions.

By now an almost incomprehensible amount of data on the properties and processing technology of PCM has been amassed. For all that, the information has been largely sporadic and has not, to this day, received comprehensive theoretical treatment, remaining mostly in the realm of empirics. Naturally, this does not undermine their value for solving purely practical problems. Search for new ways of predicting PCM behavior in case of known compositions or of developing new formulations is continuing. This tendency will most probably be a lasting one, with new applications of PCM emerging in the process of their study, where before it was deemed inexpedient or just unthinkable [12–14].

The goal which the authors of this review set themselves was to discuss, if only briefly, how the combination of the basic properties of PCM ingredients and sample preparation conditions affect the properties of the composite. To make this possible the authors had to limit themselves to just the fundamental physico-mechanical and technological characteristics of compositions with disperse and short-fiber fillers of, mainly, mineral origin. Even so, we could not dream of anything like a complete and detailed analysis of all the published material, since as early as 1981 the bibliography of works devoted to the strength of fiber-filled PCM alone counted more than one thousand titles [15] and today their number has been increased twice or more. We therefore focused our attention on those works which could prove helpful in developing ways of purposefully controlling PCM properties by different means. The means include special treatment of the filler, modification of the polymeric matrix, use of combinations of fillers, etc. The so-called polymerization or chemical filling technique of thermoplastics is very promising [16]. A number of detailed reviews on the subject have already been published, however, they laid most emphasis on the problems of synthesis and kinetics of the reactions involved [17–22]. So far there have been no attempts to undertake a comprehensive analysis of all the available data on the structure of the composite materials, their technological and physico-mechanical properties.

1 Effect of a Filler on Polymer Properties

1.1 Methods for Estimating the Filler Effect on Polymer Matrices

Conventionally fillers are divided into reinforcing, active, and inert ones. The reinforcing class includes mainly fibrous materials. Disperse fillers may also perform the reinforcing function, and then they are called active. For the criterion of activity of a filler it has been proposed to employ, for example, the extent of variation of the relative viscosity of the melt or solution caused by introduction of a filler [23] – the greater the variation the better the affinity between the polymer and the filler.

Indeed, one often observes a more or less direct relationship between the rheological properties of melts and the mechanical strength of the "condensed" material. This is a commonplace statement in regard of, say, stiffness, since the equations relating the viscosity of heterogeneous materials with their composition are formally identical with equations for the shear modulus, when the shear rate changes precisely by the value of the relative shear strain [24]; it is assumed then that the following relationship is valid [25-26]:

$$\eta_c \cdot \eta_m^{-1} = G_c \cdot G_m^{-1}, \tag{1}$$

where η_c, η_m, G_c, G_m – are the viscosities and shear moduli of the composite and matrix, respectively.

This is true when the material stiffness is much smaller than that of the filler and Poisson's ratio of the matrix is 0.5 – which is not often the case [24].

Nielsen suggested a more exact equation relating viscosity and elasticity modulus [27]:

$$\eta_c \cdot \eta_m^{-1} - 1 = 2.5(8 - 10\nu_m)\,[15(1 - \nu_m)^{-1}]\,(G_c \cdot G_m^{-1} - 1), \tag{2}$$

where ν_m is Poisson's ratio for the matrix.

From published data it may be concluded that if, upon addition of a strictly fixed quantity of different fillers to one and the same polymer matrix, the viscosity of one system is higher than that of the others one may be almost sure that this system will also exhibit better strength characteristics. This is largely due to the ability of the filler to form its own structural network ("skeleton") in the composite. On these grounds it has been proposed to use the strength of structures in model systems of the "polymer solution/filler" type measured with the help of, for instance, a plastometer with a tangential movement of the working plate [28] as a method for evaluating the filler activity and the extent to which it is affected by different surfactants and other modifiers. It is naturally assumed that the structure forming process is identical in solution, in melts and solid samples.

Yet, the increase of viscosity does not always indicate that the strength will increased symbatically. It is well known that the viscosity and moduli of solid composite based on the same matrix matrial may increase as a result of the agglomeration of filler particles [29]. At the same time, the strength of the composite, especially the impact strength, is lower in materials with agglomerates than in materials with a more uniformly distributed filler [30]. Treatment of fillers with modifying agents may also result in a situation where the system featuring a higher viscosity will have poorer strength characteristics [31, 32].

The shear viscosity, especially as measured with capillary rheometers characterized by high shear rates, is hardly sensitive to material structure since the investigator usually has to deal with the substantially destroyed structure in the molten sample. Melt stretching experiments would normally provide much more information [33].

There is greater similarity in the behavior of stretched melts and solid samples prepared by, e.g. pressure molding, probably, for the reason of parallelism in structure formation and destruction caused by deformation in melts and the amorphous regions of solid matrices. It is also possible to use identical equations for longitudinal viscosity and strength which present them as functions of the filler concentration [34].

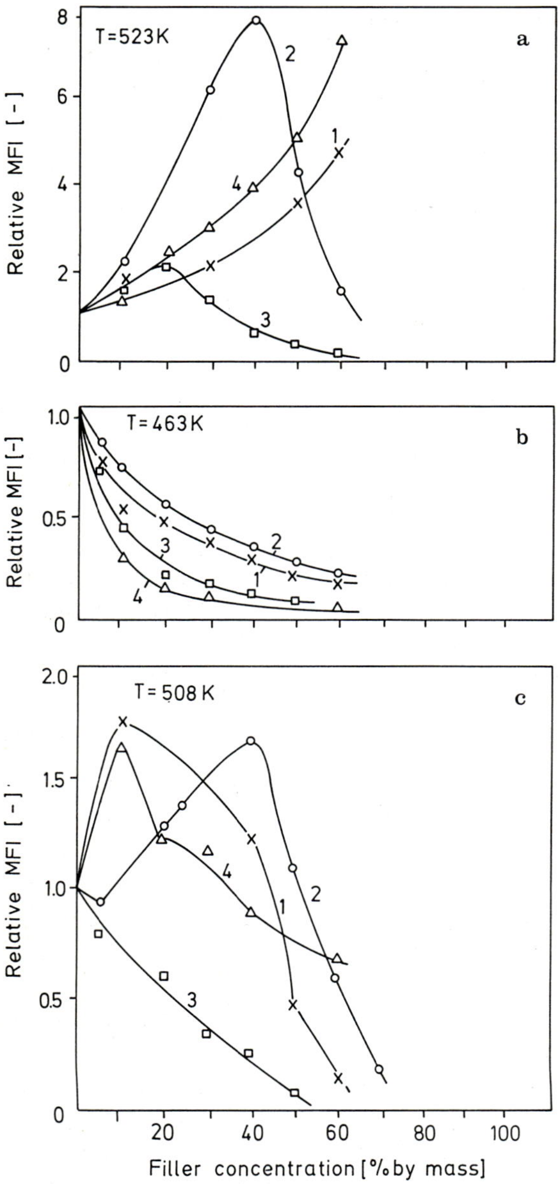

Fig. 1. Relative MFI of Diflon polycarbonate (*A*), HDPE (*B*) and polycaproamide (*C*) as function of filler concentration (φ) and type. According to the data of [35]

It should be noted that the very notion of filler activity is to a great extent conditional. Even where addition of a mineral filler to a specific polymer somewhat reduces its strength, it may increase its heat resistance, reduce the casting shrinkage, etc. Besides, a filler inert in relation to some particular group of polymers may be active in relation to another. Figure 1 illustrates an ambiguous effect of fillers upon one and the same parameter of different polymers. The figure is based on data from Ref. [35] and shows the relative melt flow index, i.e. the ratio of the

MFI of the composition to that of the matrix, as a function of the filler concentration. It can be seen that, as the concentration of a particular filler increases, the index increases too for one matrix but decreases for another, and varies by a curve with an extremum for a third one. Even for one and the same "polymerfiller" system and a fixed concentration of filler, the stress-strain characteristics, such as ultimate stress, may, depending on the testing conditions (temperature, rate of deformation, etc.) be either higher or lower than in the reference polymer sample [36].

1.2 The Interphase Concept

To be able to control the PCM properties in the desired direction it is very important to know the relationships between the material composition and properties. Since melt viscosity is one of the most important characteristics of processability of PCM, there have naturally been a large number of equations proposed for describing the viscosity versus filler concentration relationship. For the purpose of this review it may be most interesting to discuss the numerous equations which have in common the use of the value Φ representing the maximum possible volume filling by filler particles packed in one way or another, as the principal constant. Here are a few examples of such equations.

Mooney's equation [37]:

$$\eta_c \cdot \eta_m^{-1} = \exp\left[k\varphi(1 - \varphi \cdot \Phi^{-1})^{-1}\right] \tag{3}$$

Equations of the type:

$$\eta_c \cdot \eta_m^{-1} = (1 - \varphi \cdot \Phi^{-1})^{-n} \tag{4}$$

where n equals 2.5 in the Landel-Moser-Bauman equation [38] or 2 in the Maron-Pierce equation [39]. Parameter φ in these equations stands for the filler percentage in the system; parameter Φ for the limiting percentage which is achieved when volume filling is maximum. According to [40], Φ for spherical particles of one size may vary from 0.52 to 0.74, depending on the packing type. Approaches to calculation of Φ for particles of different shape have been proposed in a number of references [41–44]. The control of Φ by means of varying the filler particle size distribution allows, even for fixed φ, to change the system viscosity within sufficiently broad limits. However, the assumption that the effect of filler on the viscosity is due solely to φ and that the particle-size composition of the filler affects viscosity only through Φ, is apparently contradicted by the experimental data. Even for fillers of narrow size analysis the viscosity of filled systems depends heavily on its particle size, increasing as they decrease, as shown, for example, in [45, 46].

It is even more difficult, relying on hydrodynamical models, to explain the frequently observed decrease, rather than increase, of viscosity when a certain quantity of filler is added [47–50].

It became possible to elucidate such experimental facts when it was proposed that considerable restructuring, manifesting itself in the appearance of boundary layers, occurs in the matrix under the effect of a filler [51–56].

The boundary layers, or interphases as they are also called, form the "mesophase" with properties different from those of the bulk matrix and result from the long-range effects of the solid phase on the ambient matrix regions. Even for low-molecular liquids the effects of this kind spread to liquid layers as thick as tens or hundreds or Ångström [57, 58]. As a result the liquid layers at interphases acquire properties different from properties in the bulk, e.g., higher shear strength, modified thermophysical characteristics, etc. [58, 59]. The transition from the properties prevalent in the boundary layers to those in the bulk may be sharp enough and very similar in a way to the first-order phase transition [59].

The thickness of interphases in polymer melts and solutions may be much greater than in low-molecular matrices owing to the greater size of their macromolecules and existence in them of submolecular associates [51].

The fraction of polymer confined in the interphase layers was estimated using the empirical formula [60, 61]:

$$\nu_{\mathrm{i}} = 1 - \Delta C_{\mathrm{p_c}} \cdot \Delta C_{\mathrm{p_m}}^{-1}, \tag{5}$$

where ν_{i} is interphase volume; $\Delta C_{\mathrm{p_c}}$ and $\Delta C_{\mathrm{p_m}}$ are sudden increments of heat capacity at glass transition for the composite and the matrix, respectively.

Rheological methods of measuring the interphase thickness have become very popular in science [50, 62–71]. Usually they use the viscosity versus concentration relationships in the form proposed by Einstein for the purpose [62–66]. The factor K_0 in Einstein's equation typical of particles of a given shape is evaluated from measurements of dispersion of the filler in question in a low-molecular liquid [61, 62], e.g., in transformer oil [61]. Then the viscosity of a suspension of the same filler in a polymer melt or solution is determined, the value of K_{eff} is obtained, and the adsorbed layer thickness is calculated by this formula [61, 63, 64]:

$$\delta = 0.5\mathrm{d}(\sqrt{K_{\mathrm{eff}} \cdot K_0^{-1}} - 1), \tag{6}$$

where d is particle diameter.

In case of a polydisperse filler with a specific surface area S_{spec}, it has been proposed to calculate the "mesophase" thickness the product $\Delta\varphi_{\mathrm{eff}} \cdot S_{\mathrm{spec}}^{-1}$ (here $\Delta\varphi_{\mathrm{eff}}$ is the effective increase of filler volume per unit of its mass) [50, 65].

The majority of investigators consider it permissible and convenient to use, when calculating the boundary layer thickness, the relationships describing the concentration dependence of viscosity in the high and medium concentration range (basically Mooney's equation) [67–71].

Kuznetsov et al.'s methodological approach [72–75] provides another example of attempts to evalue the interphase thickness experimentally. Their approach was based on the hypothesis that the mesophase remains glassy while the bulk of the binder has already passed to the highly elastic state. Investigating the concentration dependence of the dynamic modulus of the filled polymer in the highly elastic and glassy states and extrapolating these relationships to intersection, the authors obtained the so-called critical filler concentration. Introducing, further, the conception of reduced filler concentration, φ_{red}, equal to the actual/critical filler concentration, they came up with relationships between a number of physico-mechanical and thermo-physical parameters of the system and the reduced filler concentration, invariant relative to the nature of filler and matrix.

Other methods for estimating the volume percentage of the interphases in a composition have been proposed, too, for example, measurements of density variations [76, 77], volume of compressed sample [78], the dielectric constant [77], etc. The important thing is that the interphase thickness determined in one way or another is an effective value dependent upon the conditions and type of the experiment by which it was determined [51].

Today it has been demonstrated convincingly that on the solid-polymer boundary there exist the adsorbed layers of polymer characterized by a density greater than that of the unfilled material [52–54, 79–82]. However, the phenomenon of physical ageing of filled polymers noted in a number of investigations (e.g. in Ref [83]) cannot be attributed to condensation of boundary layers alone. The phenomenon itself consists of a slow increase in the density of a filled system during annealing, the variation being slower the higher the adhesion between the polymer and filler and the greater the filler concentration. This means that a considerable free volume is formed during the filling operation which gradually relaxes. Indeed, apart from condensation of a certain quantity of polymer some investigators noted the lossening effect of the filler upon a portion of the matrix [52–54, 84–86]. According to Lipatov [54] denser volumes can be expected to form on the filler surface and, where the adhesion forces are weaker than cohesion, the mass transfer must by directed to the surface, since its "force field" helps to keep the macromolecules fixed. Condensation of the polymer on the surface is an inevitable result of the polymer transport to the interphase regions, whereby layers are inevitably generated, alongside the denser layers, where all the mobile macromolecular segments have moved into the condensed boundary layer, whereas the macromolecular segments that are not sufficiently free to move about from the loose zone. Where cohesion energy becomes greater than the adhesion energy the loose layers are, on the contrary, formed directly on the filler surface. The author considers the loose packing as the analog of the unordered inter-structural portion of the bulk polymer. The resulting structure is nonequilibrium and will relax in time. It is suggested that we consider these structures as regions consisting of virtual phase particles – phasones – and treat them as phasone-like systems [54, 87].

Rozenberg and Irzhak [88, 89] advanced a somewhat different hypothesis concerning the manner in which the structure of the filler polymer matrix might be formed. In their opinion, the long-range full-scale condensation effect is largely due to the local fluctuations of matrix density owing to the differences in thermal expansion coefficients, and in reacting thermoreactive systems also due to the so-called chemically-induced shrinkage. Note that matrix defects may indeed increase considerably with increasing filler concentration [90]. As suggested by the authors of [88, 89], the generated excessive free volume may diffuse through the matrix either remaining in original dimensions or breaking down into smaller regions, or even merging into larger ones. The diffusion of "holes" leads to condensation. Where adhesion is low or absent the polymer-filler boundary will be absorbing and the free volume will accumulate at the interphase, whereby the binder may come off the filler surfaces. In case of high adhesion the free volume may accumulate some distance away from the surface, generating a nonmonotonous distribution in the direction normal to the particle surface. If the interaction is nonuniform across the interphase, regions with locally increased or decreased density may arise in the boundary layer of even one individual particle. When the excessive free volume does emerge to the surface, the matrix undergoes a kind of multilateral stretching, that is, considerable internal stresses are developed within it.

Lipatov [54] as well as Rozenberg and Irzhak [88, 89] believe that different impurities may diffuse to the rarefied zones, whereby the matrix becomes more ordered throughout its volume.

It is precisely the loosening of a portion of polymer to which the authors of [47] attribute the observed decrease of viscosity when small quantities of filler are added. In their opinion, the filler particles added to the polymer melt tend to form a double shell (the inner one characterized by high density and a looser outer one) around themselves. The viscosity diminishes until so much filler is added that the entire polymer gets involved in the boundary layer. On further increase of filler content, the boundary layers on the new particles will be formed on account of the already loosened regions of the polymeric matrix. Finally, the layers on all particles become dense and the viscosity rises sharply; after that the particle with adsorbed polymer will exhibit the usual hydrodynamic drag.

The results of [91] supply ample evidence in support of this qualitative picture. The authors determined the baric viscosity factor $b = [\partial \ln \eta (\partial P)^{-1}]\, T$ (where η is viscosity, P is pressure) for impact-resistant polystyrene filled with antimony trioxide. The viscosity piezocoefficient is known to be related to the free volume. A very simple formula for this relationship has been proposed in [92] in this form:

$$b = V_h(RT)^{-1}, \tag{7}$$

where V_h is volume of holes (the free volume); R is the gas constant; T is absolute temperature.

The authors of [91] observed the viscosity going down to the minimum when the SB_2O_3 powder concentration was about 1%, whereas b reached a sharp peak under these conditions.

The interphase composition depends on the cohesion/adhesion energy ratio. The higher molecular polymer fractions are usually adsorbed on the high-energy surfaces of mineral fillers, the molecular mass gradient extending as far as a maximum of 20 microns [93]. Across a boundary layer of this kind there must be a gradient of, actually, any MM or MMD-dependent property. Vice versa, on a low-energy surface such as that of polymeric reinforcement fibers, the interphases are rich in low-molecular fractions [94]. In case of a porous filler, be it with a high surface energy, one may expect selective sorption of those macromolecules whose size is commensurate with pore size. The smaller the pore diameter, the lower the molecular mass of the polymer absorbed, and the coverage factor of the porous surface is the higher the more fractions with a low molecular mass are available in the mixture [95]. On one and the same filler the adsorbed layer thickness linearly decreases with the mean M_w of the matrix [96]. The interphase layer thickness in a mixture of polymers is higher than in each of its components, the layers being rich in macromolecules of one of the polymers [97]. The polymer volume in the boundary layer at fixed filler concentration is larger when the macromolecular flexibility is lower [98]. Higher stiffness of macromolecules also entails an increased proportion of excessive free volume.

The authors of [99] proposed a calorimetric method for determining the degree of the polymer-filler interaction: the exothermal effect manifests itself in the high energy of the polymer-filler adhesion, the endothermal effect is indicative of a poor, if any, adhesion. The method was used to assess the strength of the PVC-Aerosil interaction with Aerosil surface subjected to different pre-treatments [100]. Comparison of the results reported in this work with the data obtained in [101] shows that the interphase thickness in the system in question increased with the strength of the polymer-filler interaction.

On the basis of the interphase conception, Lipatov and Privalko proposed in [102] to consider as the lower limit of existence of high-filler polymeric materials such a filler concentration for which the boundary layers overlap. This critical concentration may be determined, for example, according to the sudden change of heat capacity at the glass-transition point (cf. [60]). The authors suggest that the maximum concentration should not be higher than that for which the interparticle layer thickness becomes commensurate with $2\langle R_g \rangle$ where R_g is the average inertia radius of the molecular ball. A further decrease of the layer thickness should result in a disturbance of the thermodynamic stability of the system as a consequence of the sharp limitation of the conformational entropy of macromolecules.

1.3 Crystallization and Melting of Polymers in Interphases

A considerable structural inhomogeneity has also been observed to be caused by a filler introduced into a matrix in the process of crystallization. In this connection the so-called trans-crystallite structures typically produced by fibrous fillers may be of interest. The phenomenon consists of the formation (around the fibers introduced into a crystalline matrix) of rather special shells made up of columnar elements whose shape is totally different from any structures occurring in the matrix [103, 104] This behavior is apparently due to the high density of the active sites on the fiber surface. Multiple nucleation hampers spherolite growth on the sides and transforms them into column-like structures. The trans-crystallite shell thickness is sensitive to melt viscosity and decreases when the latter increases. It may be much greater than the filler fiber diameter [104]. Interestingly, the crystallization develops in this way almost exclusively with organic and carbon fibers, whereas no such behavior was observed in systems with untreated glass

Table 1. Effect of filler nature, dispersedness and concentration on the degree of crystallinity of polyisobutylene samples [105]

Filler	Mean particle diameter, m	Filler concentration, p.b.v.	Degree of crystallinity %
$MgCO_3$	0.06	10	48
$MgCO_3$	0.06	20	40
MgO	0.12	10	54
MgO	0.12	20	51
MgO	0.12	30	28
BaO	0.165	10	12
Rechnical carbon	0.09	20	59
PM-30	0.09	20	65
Polypropylene	7.3	10	44
powder	7.3	20	48
powder	7.3	50	59
Unfilled poly-isobutylene	—	0	44

fibers except in a very few cases. The trans-crystallite structures arising in a system affect some or most of the physico-mechanical characteristics of the composition [103].

In dispersed systems the nature of the filler also plays a controlling role in the way the crystallization proceeds. Examples are reported in [105], whose authors have used X-ray analysis to estimate the degree of crystallinity of polyisobutylene filled with different concentrations of a number of filler materials, after 100 cycles of 50% stretching. Polyisobutylene crystallizes as a result of such treatment. The results are given in Table 1.

From the table data one sees that, again, it is carbon or polymeric filler which provides for the maximum degree of crystallinity.

Malinsky et al. [106] investigated the effect of the free surface energy of the substrate on spherulite growth rate in polypropylene. They found that the spherulite growth rate decreased drastically as the free surface energy of the substrate increased, no matter what the state of aggregation of the substrate. The structure of the crystalline formations in interphases becomes different from that in the bulk of material. The melting point of polypropylene crystallized as a block is 14% lower than the melting point of the same material crystallized as a thin film upon a glass plate [107]. The authors of [108] reported an even greater difference between the melting points of material at interphases and in bulk. Note that normally the variation of the specific heat of melting is directly proportional to the variation of matrix density [109, 110].

Mamuni et al. [110] have shown that modification of the filler may also have a marked effect on structurization in the interphases. They measured the variation of the specific heat of melting (ΔH) of low- (LDPE) and high-density polyethylenes (HDPE) filled with glass powder with modified and unmodified surface. The modifying agent was microdiisocyanate prepared from poly(ethylene glycol adipinate) and toluylene diisocyanate. The low-molecular modifying agent applied to the glass surface reacted, later in the process, with atmospheric moisture to become resilient poly(urethane urea) which adheres perfectly to both the filler and the matrix. The modifier layer thickness was varied from 0.001 to 0.2 microns.

In case of HDPE, even the tiniest layer of poly(urethane urea) reduces ΔH, which may be attributed to inhibition of crystallization. Unexpectedly a further increase of the layer thickness brings about a sharp decrease of H, apparently, due to the inhibited interaction in the base-polymer – filler system. From the modifier layer thickness of about 0.01 microns on the value of ΔH decreases again, probably, on account of the formation of extended boundary polymer-polymer layers as suggested in Kuleznev's monograph [111]. The manner of base polymer density variation with the increase of modifier layer thickness was found to besimilar to the variation of ΔH. Morphological analysis has shown that unmodified filler has an insignificant effect on the HDPE structure; modification of the filler disturbed the degree of ordering of submolecular structure.

The variation of mechanical parameters correlates with variation of ΔH: sudden drop of tensile and impact strength with a thin modifier layer, then increase of strength with increase of poly(urethane urea) film – a minor increase of strength but a tangible one for specific impact strength.

The situation with LDPE is totally different. Addition of unmodified filler dramatically decreases the specific heat of melting. Provision of a film of modifying agent somewhat increases it. With a poly(urethane urea) film thickness of 0.03 microns the increase of ΔH discontinues and then, as the film thickness goes on increasing, the ΔH value decrease monotonously. The density, in contrast, increases in the entire range of modifier film thicknesses. Tensile strength and relative elongation at rupture are sharply increased if a thin modifier film is provided; as the thickness increases, the parameters smoothly decrease.

We have given so much attention to the results of [110] because they present a very convincing illustration of some effects resulting from different modifications of the filler. The fact that is particularly remarkable is that even for polymers of a very congenial nature (LDPE and HDPE) the response to filler may be widely different. In this light, one is hardly surprised at the conclusions of the authors of [78] concerning the considerable increase of the degree of crystallinity at polyethylene interphases in matrices filled with molybdenum powder. Another important conclusion was that the sharply enhanced interaction in the polymer-filler system does not, in all cases, lead to improved physico-mechanical characteristics. The situation where the modification may, instead, result in a deterioration of properties of the composite if only because of polymer amorphization, is not at all improbable. Similar behavior may also result from the formation of hydrogen bonds [112]. Formation of chemical bonds should enhance the amorphization process [113]. As shown in [114], surface finishing may modify the structure of a polymer sample. The authors investigated PP crystallization between two mica plates. They observed molecular chain orientation in a direction perpendicular to the mica surface. After mica treatment with chlorinated hydrocarbon or silane they observed disorientation in boundary layers and decrease of the degree of crystallinity.

1.4 Effect of Interphases on Physico-Mechanical Properties

Selective sorption by the filler sorption of one of the matrix components (residual monomer, low molecular homologs, various impurities) may lead to plasticization of the boundary layers and appearance of "soft" interphases [115, 116].

Macromolecular fixation on the filler surface, their orientation in the direction complanar or normal to the surface and condensation result in "hard" interphases [116].

It cannot be put to question that interphases should have a certain effect on the final composite properties. It is however impossible with the available means to investigate the effect of the properties and thickness of the interphases in the pure form. First, to provide a homogeneous film on the filler with a controlled thickness and properties is no easy task in itself. Second, addition of the modified filler to polymer produces, as shown in, e.g. [110], a system with a real sequence of interphases, viz.: the filler-sublayer interphase, the sublayer-base polymer interphase. This resulted in the creation of theoretical models allowing for the degree of heterogeneity in transition from the filler to bulk polymer. However, as noted in [117], in the majority of models proposed for describing the composition/propert relationships the polymer filler interaction is allowed for with, at best, the help of an empirical coefficient. This approach can be useful for estimating the filler effect on the viscosity of filled systems. It is usual to assume, when estimating viscosity, that the formation of interphases entails an increase in the effective volume of particles. When using equations for predicting viscosity in the medium and high filler concentration ranges, for example, Mooney's equation [37], this may effect the value of Φ (the maximum filling ratio) and, possibly, Einstein's coefficient – K. From the experimental data one finds the constants for the selected equation, the interphase thickness and concludes about the extent of the filler-matrix interaction [67–71]. In some cases, however, the results one obtains do not have any physical meaning.

The method used for predicting the different physical and mechanical characteristics of crystalline or glassy polymeric composites is somewhat different. Most frequently it has been proposed (cf., e.g. [118]) to introduce an extra term into the relationship between some characteristic and the composition of the material based, more often than not, on the principle of additivity of the filler and matrix characteristics and taking into account their relative volumes in the composite. This extra term is the product of the interphase volume by a characteristic other than the characteristics of either the matrix or the filler.

In this conceptual framework it is naturally impossible to simulate the effect of the interphases of complex structure on the composite properties. A different approach was proposed in [119–123]. For fiber-filled systems the authors suggest a model including as its element a fiber coated with an infinite number of cylinders of radius r and thickness d_r, each having a modulus E_r of its own, defined by the following equation:

$$E_r = E_m + E_f(r_f \cdot r^{-1})^{2K_1} - E_m(r_f \cdot r^{-1})^{2K_2}, \quad (8)$$

where E_m, E_f are the moduli of the matrix and the fiber, respectively; r_f is fiber radius.

The exponentials K_1 and K_2 in the model account for the degree of heterogeneity of the interphase layer. The ratio of these exponentials is representative, as suggested in [121], of the load transmission between the fiber and the matrix (the adhesion factor α).

When $\alpha \to 1$ ("perfect" adhesion) the elasticity modulus of the interphase decreases continuously from the fiber value to the matrix value, the interphase layer modulus being higher than that of the matrix. When $\alpha < 1$, the interphase layer modulus assumes, some distance off the fiber surface, a minimum value smaller than the matrix value, and then increases tending asymptotically to the matrix modulus.

Lipatov et al. [116, 124–127] who simulated the polymeric composite behavior with a view to estimate the effect of the interphase characteristics on composite properties preferred to break the problem up into two parts. First they considered a polymer-polymer composition. The viscoelastic properties of different polymers are different. One of the polymers was represented by a cube with side a, the second polymer (the binder) coated the cube as a homogeneous film of thickness d. The concentration of d-thick layers is proportional to the specific surface area of cubes with side a, that is, the thickness d remains constant while the length of the side may vary. The calculation is based on the Takayanagi model [128]. From geometric considerations the parameters of the Takayanagi model are related with the cube side and film thickness by the formulas:

$$a(a + 2d)^{-1} = \Phi\,, \qquad \lambda = \Phi^2\,. \tag{9}$$

In calculation the authors of the model assume that the cube material possesses the complex modulus E_A^* and mechanical loss tangent tg δ_A which are functions of temperature T. The layer of thickness d is composed of material characterized by a complex modulus $E_B^* = f(T + \Delta T)$ and tg $\delta_B = f(T + \Delta T)$. The temperature dependences of E_B^* and tg δ_B are similar to those of E_A^* and tg δ_A, but are shifted towards higher or lower temperatures by a preset value ΔT which is equivalent to the change of the glass transition point. By prescibing the structural parameters a and d one simulates the dimensions of the inclusions and the interlayers, and by varying ΔT one can imitate the relationship between their respective mechanical parameters.

The model for a filled system is different. The filler is, as before, represented by a cube with side a. The cube is coated with a polymer film of thickness d; it is assumed that d is independent of the filler concentration. The filler modulus is much higher than that of the d-thick coat. A third layer of thickness c overlies the previous one and simulates the polymeric matrix. The characteristics of the layers d and c are prescribed as before, and the calculation is carried out in two steps: at first, the characteristics of the filler (a) – interphase (d) system are calculated; then this system is treated as an integral whole and, again, as part of the two component system (filler + interphase) – matrix. From geometric considerations the dispersed filler concentration is given by:

$$V_f = a^3[(a + 2d + 2c)^3]^{-1} . \tag{10}$$

The Takayanagi model parameters are related with filler concentration and interphase thickness by the following simple relationships:

$$\begin{aligned} \Phi_1 &= a(a + 2d)^{-1}; \qquad \lambda_1 = \Phi_1^2 , \\ \Phi &= (a + 2d + 2c)\,(a + 2d)^{-1}; \qquad \lambda_2 = \Phi_2^2 . \end{aligned} \tag{11}$$

For given values of a, d and V_f, the value of c is determined from the formula for V_f. Analysis of the data obtained by the above procedure has permitted the authors of the quoted references to make a number of qualitative conclusions:

– in order for the dependence of the mechanical characteristics of filled polymer on filler particle size to be manifest it is necessary that the mechanical properties of the interphase be markedly different from those of the basic phases – the greater the difference, the stronger the dependence.

– if the interphase has a higher modulus than the matrix, the modulus and mechanical strength of the composition will increase inversely proportionally to the particle size;

– for composites with "soft" interphases the decrease of filler particle size should manifest itself in decrease of elasticity and strength in accordance with a law stronger than simple inverse proportionality;

– the greater the filler concentration and interphase layer thickness, the more tangible is the effect of particle size on composite properties;

– one common point of composites with "soft" and "hard" interphases is that they both have an upper and a lower limit of elasticity and strength due to the corresponding limits of the interphase and the matrix;

– if the interphases and the matrix are in the glassy state, the interphase will have almost no effect on the composite strength; in the transition region the "soft" interphases increase and "hard" interphases decrease the composite strength.

It should be noted once more that filled polymers may be treated as three-phase systems only nominally, since thermodynamically the interphase cannot be regarded as a phase in its own right.

The above qualitative conclusions made on the basis of the results of [116, 124–127] correlate with the results of [129, 130] in which the calculation is based on composite models with "nucleus-shell" inclusions. The authors illustrate this with the calculation of a system consisting of a hard nucleus and elastomeric shell in a matrix of intermediate properties, and a system where the nucleus and matrix properties are identical whereas the shell is much more rigid. The method may, however, be also applied to systems with inclusions where the nucleus is enclosed in a multi layer shell. Another, rather unexpected, result follows from [129, 130]: for a fixed inclusions concentration, the relative modulus of the system decreases with increasing nucleus radius/inclusion radius ratio, that is with decreasing shell thickness.

1.5 Effect of Filler Particle Size and Shape Upon the Physico-Mechanical Properties of Composites

Solomatov et al. [131] derived an equation for strength at the polymer – spherical particle interface:

$$\sigma_p = \sigma_\infty \exp\,[W_K(KT)^{-1}] \exp\,[2W_S V(d_K T)^{-1}]\,, \tag{12}$$

where W_S is the fraction of specific surface energy per one macromolecule; W_K is the work of cohesion calculated per macromolecule; σ_∞ is the ultimate stress in the zone of adhesion contact between the substrate and matrix over a plane surface; d is mean particle diameter; K is Boltzmann's constant; T is absolute temperature; V is the volume of a one macromolecule of the adhesive layer.

From Eq. (12) it follows that when the reciprocal of mean particle size increases, the value of σ_p must increase e times. Indeed, Alter [132] showed that some of the physico-mechanical characteristics are a linear function of the logarithm of mean particle size. Physico-mechanical properties are better when the particle diameter is smaller. At the same time, we can think, on the basis of the discussion in the previous section, of a situation, noted in [133], where there really is some increase of the modulus of the filled polymer with concentration of microspheres, the relationship being the stronger the smaller the filler particle size. Whereas the strength remained practically unchanged whatever the glass sphere size. But while strength does not always show a great dependence on particle size, impact strength may vary within very broad limits when the filler particle size is varied [134–137]. For example, for finely dispersed filler the specific impact strength of polymeric composites may increase with filler concentration [133, 134] (sometimes markedly so), but with larger particles this parameter drops dramatically [135, 136]. This behavior of composites is conventionally attributed to the enhanced energy absorption as a result of the longer path which the crack has to go through because it has to bypass the numerous tiny particles, whereas larger particles play the role of stress concentrators. Having noted that the smaller the particle size of dispersed filler for its fixed concentration the higher the strength of the composite, the authors of [90] show that, first, the strength of the composite with dispersed particles is lower than that of the matrix, and second, that the strength versus particle size curve gradually tends to a limit, that is, upwards of a certain size it stops being dependent on the effective particle diameter (d). However, in a series of studies Japanese investigators refuted these premature conclusions [138–141]. The authors studied the effect of ultra-dispersed filler on the yield strength σ_y of some thermoplastics. The matrix materials were polypropylene [138], polycaproamide [139], poly(methyl methacrylate) [140], poly(vinyl chloride) [141]. The filler was silica with mean particle sizes of 70, 160 and 400 Å, and with particles of 30 microns and larger. They investigated the effect of stretching rate, temperature, and volume content of filler and showed that both σ_y and stress at the yield strength plateau were, indeed, lower in samples with particles in the micron size range than in the matrix, but were higher than in the matrix in samples with ultradisperse filler. The σ_y values linearly increase with deformation rate logarithm ($\dot{\varepsilon}$), the slope of $\sigma_y = f(\dot{\varepsilon})$ decreasing with temperature. The yield strength versus effective particle diameter (d) relationship increases for all matrices almost linearly, in the double logarithm coordinates, as the filler particle size decreases (dispersity increases) and shows no tendency for saturation in the entire investigated d range. The authors concluded that the reinforcing effect of one and the same type of filler changes in different polymeric matrices in function of the surface tension of the polymer in such a way that it is greater when the surface tension is greater.

In passing from spherical particles to particles of an anisodiametrical shape (ellipsoids or fibers) the stress resisted by the filler is the higher the more pronounced the anisodiametricity of the particles [142].

Chow demonstrated theoretically [143] that for anisodiametrical particles, the ultimate tensile stress is inversely proportional to square root of the effective or characteristic filler particle size (in this case by effective particle size the ratio of particle volume to surface area is implied).

The behavior of the physico-mechanical characteristics of polymeric composites is easily traceable in the table given in [144] which presents the results of experiments with polyamide matrices filled with resite particles of different shape. The filler concentrations were adjusted so that the integral contact surface area in the filler-matrix system remained the same.

The data given in Table 2 show that fibrous fillers have the optimum shape in terms of the maximum reinforcing effect, as could be expected.

Well known are Kelly and Tuson's [145, 146] formulas for the strength of fiber-filled composites:

$$\sigma_c = V_f[1 - l_c(2l)^{-1}]\,\sigma_f + (1 - V_f)\,\sigma_m \quad \text{for} \quad l \geqq l_c\,, \tag{13}$$

$$\sigma_c = V_f \cdot \tau_m \cdot l \cdot d^{-1} + (1 - V_f)\,\sigma_m \quad \text{for} \quad l < l_c\,, \tag{14}$$

where V_f, σ_f are the content by volume and strength of the fiber; σ_m is stress in the matrix under deformation which breaks the fiber; τ_m is shear stress at the fiber-matrix interphase; l and d are the length and diameter of fiber respectively.

Table 2. Physico-mechanical properties of polyamide matrices filled with resite particles [141]

Parameter	Filler particle shape and filling ratio % by vol.		
	Spheres, 20%	Random, 7%	Fiber, 5%
Specific breaking stress, $\times 10^{-5}$ cm:			
– tensile	5.4	5.9	6.8
– compressive	8.7	9.0	12.6
– bending	10.0	10.8	4.1
Specific modulus of elasticity in compression, $\times$ 10 cm	3.4	3.6	4.1

The parameter l_c, the critical fiber length, is given by:

$$l_c = \sigma_f \cdot d(2\tau_m)^{-1} . \quad (15)$$

In other words, the critical fiber length is a function of its strength, diameter and the shear strength at the filler-matrix interface.

Ternopolsky and Skudra [147] believe that the calculated l_c value must be increased by the doubled length of a certain friction segment, since l_c represents only a portion of the fiber length affected by tangential stresses. The underlying idea of this recommendation is that the breaking apart of the adhesion joints between the fiber ends and polymeric matrix is inevitable, because adhesion forces are, like the shear strength of the polymeric binder, much smaller than the strength of the fiber itself. Transmission of forces from the matrix to the fiber is possible if additional forces arise at fiber ends which prevent their displacement relative to the matrix. The authors postulate two "transmission" mechanisms in orientated fiber-filled composites, viz. an adhesion mechanism (until the adhesion bonds are broken at fiber ends) and the friction mechanism (after the shear).

What distinguishes the fibers used as reinforcements is the so-called scale effect of strength which manifests itself in strength dependence on the "contact" or "cramp" length. In [148, 149] it was shown that the reinforcing fiber strength distribution function is approximated with sufficient accuracy by the Weibull distribution, although there is some degree of polymodality. The average strength distribution curves (in the double logarithm coordinates) linearly depend on the "cramped" length of the fiber, decreasing as the latter increases.

In [148] it was shown that the strength of the composite depends on the fiber strength at the so-called "ineffective" length, so that only a few percent of the total amount of the ineffective fiber lengths in the composite become broken before the specimen fails completely. The authors of [148] proposed to express the relationship between the average fiber strength ($\langle\sigma_f\rangle$) on the "cramped" length l in the form:

$$\lg \langle\sigma_f\rangle = A - B \lg (l \cdot l_0^{-1}) , \quad (16)$$

where the constants A and B are related with Weibull distribution parameters; l_0 is a certain standard length; l is specimen length. The ways by which A and B are found are described below [149, 150].

Under tensile stresses, the fibers in a composite specimen are broken exactly where there are the most critical surface defects. The defects that may occur on the surfaces of the resultant fragments cannot be as critical, because the specimen has already failed at these points. The authors of [150] emphasize that this is precisely the difference between the failure of composites and fiber testing for different "cramped" lengths where the reduction of fiber length can only reduce the probability of occurrence of dangerous defects but cannot bring this probability to zero. Therefore, the fiber strength distribution may be different in fiber testing at different "cramped" lengths and in their crushing in the composites, and its approximation with Weibull distribution cannot be used except with reservation [150].

As fibers are crushed in the composite, their strength increases. So long as the fibers interact with each other they can stand a load. When, however, the fiber length becomes so small that the shear stresses between the fiber and the matrix become about as low as the shear strength at the interphase the crushing process will stop and fiber fragments will be pulled out of the matrix.

The critical and noneffective fiber length can be found by the Kelly-Tyson formula in the form:

$$l_c = \langle \sigma_f \rangle_{l_c} d_f (2\tau)^{-1} , \tag{17}$$

where $\langle \sigma_f \rangle_{l_c}$ is mean fiber strength as length l_c.

Since fibers cannot be comminuted to fragments shorter than l_c, it was proposed in [149] to call the parameter $\langle \sigma_f \rangle_{l_c}$ the ultimate realizable (attainable) strength (σ_{lim}) and compare it with σ_{cr} when assessing the composite properties.

Equations were obtained in [150] relating σ_{lim} and the critical length with fiber diameter, adhesion to the matrix (expressed in terms of shear strength) and Weibull distribution parameters:

$$\lg \sigma_{lim} = 1(1 + B)^{-1} (A - B \lg d_f + B \lg 2\tau) , \tag{18}$$

$$\lg l_c = 1(1 + B)^{-1} (A + \lg d_f - \lg 2\tau) . \tag{19}$$

Analysis of the above formulas leads one to conclusion that ultimate attainable strength of material increases with the increasing scatter of the strength of fine fibers characterized by good adhesion to the matrix and is lower for fibers with larger diameters and poorer adhesion.

However, in composites, σ_{lim} is hardly ever attained, since fiber ruptures result in development of mainline cracks even at lower stresses. It is therefore necessary to know the so-called attainment coefficient in order to predict the mechanical properties of composites. The fiber strength in a PCM $\sigma_p = k\sigma_{lim}$ (here k is the attainment coefficient), and Eq. (18) becomes:

$$\lg \sigma_p - \lg k = 1(1 + B)^{-1} (A - B \lg d_f + \lg 2\tau) . \tag{20}$$

Parameters A, B, k and d_f for PCM may be determined in several ways [150] which are briefly discussed below.

1) From the results of fiber testing for different "cramped" lengths one determines A and B (cf. [147]). The composite containing this particular fibrous filler is then tested under tension or bending to determine τ. Then the attained strength is compared with the limiting one calculated by Eq. (18) and the coefficient k is found; then one can predict k for any τ value.

2) The fiber strength is determined for a particular "cramped" length (for example, for the 10 mm base) and the specimen is tested under tension (bending) at two P values. One then solves the set of simultaneous equations:

$$\lg \langle \sigma_{10} \rangle A - B\,, \tag{21}$$

$$\lg \sigma_{p_1} - \lg k = 1(1 + B)^{-1} (A - Bd_f + \lg 2\tau_1)\,, \tag{22}$$

$$\lg \sigma_{p_2} - \lg k = 1(1 + B)^{-1} (A + Bd_f + \lg 2\tau_2)\,. \tag{23}$$

It is assumed that k and σ_p depend on the filling ratio (V_f), fiber modulus, method of fabrication. In [151] for example, the k for carbon fiber bundles has been found to be:

under tension

$$k_{tens} = 1 - B(0.12E_f + 3.8V_f - 2.9)\,, \tag{24}$$

under bending

$$k_{fab} = k_{tens} \cdot 0.81^{-1}\,. \tag{25}$$

In broken fiber-filled composites there have been observed fiber fragments much shorter than the critical length [152]. The reson why brittle fibers undergo fragmentation has been attributed [152] to stress surges arising in the failing fiber and propagating along it. Consecutive fragmentation of fibers into approximately equal parts is due to interference of the loading wayes propagating from the failure sites. The probability of further ruptures decreases after separation of fibers from the matrix begins, whereby the stress surges are hindered.

The strength characteristics of composites with an anisodiametric filler depend heavily on the orientation of the reinforcements relative to the direction of load application. For fiber-filled materials this has been observed in a number of works (cf., e.g. [133, 153–157]); illustration can be borrowed from [157] the data of which are presented as Table 3 below.

Table 3. Breaking stresses for polypropylene with 30% of glass fibers in function of the angle between the fiber orientation and direction of load application [157]

Angle between fiber orientation and direction of load application, degrees	σ_{tens}, MPa
Unfilled PP	14.62
0°	67.6
15°	52.3
25°	50.4
35°	40.9
45°	38.6

Apart from the conditions of load transmission from fiber to matrix, the anisotropy of mechanical characteristics is also due to the considerable anisotropy of the fibers themselves in the longitudinal and transverse directions, especially in the case of fibrous reinforcements of polymeric nature [154].

The effect of orientation on the physic-mechanical properties is taken into account with the help of appropriate coefficients in the respective formulas [152, 155, 157]. The procedures for determining these coefficients are described in the same references.

The contribution of the orientation of scaly fillers to the complex of composite properties is no smaller; but in contrast to fibers, the scales can reinforce the composite in two directions [158].

The modulus of a composite filled with scales of some material is usually determined with the help of the so-called mixture rule [159]:

$$E_c = kE_fV_f + E_mV_m \,, \tag{25}$$

where k is a coefficient which takes into account the length and degree of overlap of the scales and tends to 1 for very long scales and a high degree of overlap; for smaller scales $k < 1$.

In accordance with the data reported in [159], there have been several equations proposed for the calculation of k, for example, Raylie's equation

$$k = 1 - \ln(U + 1)\,U^{-1} \,, \tag{26}$$

where

$$U = \alpha[G_m \cdot V_f(E_fV_m)^{-1}] \tag{27}$$

here, G_m is the shear modulus of the matrix; α is some characteristic ratio for given filler particles.

The breaking stress may be evaluated using the formula proposed by Pegeaut [159]:

$$\sigma_x = k'\sigma_f[1 - \sigma_f(\sigma_\alpha \cdot \tau_m)^{-1}]\,V_f + \sigma'_mV_m \,, \tag{28}$$

where σ_x is the breaking stress under tension directed along the orientation; α is the mean characteristic ratio for the scales; τ_m is shear strength of matrix. The value of k' depends on the packing pattern of the scales and varies from 1/2 for symmetrical packing to 2/3 for the staggered packing.

1.6 Effect of Filler Particle Size and Shape on the Rheological Properties of Composites

Particle size and shape have a strong effect on the rheological properties of materials as well, viscosity among them.

As an illustration we refer to the data of Lackdavala and Salovey [160] who studied the flow of styrene/butyl methacrylate copolymer composites with ash of

Table 4. Effect of specific surface area of filler and its concentration on concentration shift factors a_{c_l} and a_{c_s} for composites copolymer + ash (the reference concentration 10%)
(a)*

Filler content, %	a_{c_l} values for ash with specific surface areas, m^2/g			
	24	55	525	625
5	5.16	5.45	6.25	6.51
10	1	1	1	1
20	3.27×10^{-1}	2.68×10^{-1}	1.66×10^{-1}	1.25×10^{-1}
30	4.36×10^{-2}	3.63×10^{-2}	2.03×10^{-2}	6.00×10^{-3}

* Reduced to concentration of 10%

(b)*

Specific surface area, m^2/g	a_{c_s} values for filler concentration of (%):			
	5	10	20	30
24	2.35	3.91	4.20	4.35
95	1	1	1	1
525	4.20×10^{-1}	2.95×10^{-1}	1.66×10^{-1}	1.25×10^{-1}
625	4.36×10^{-2}	3.63×10^{-2}	2.03×10^{-2}	6.00×10^{-2}

* Reduced to $S = 95\ m^2/g$

different particle size (specific surface area). The authors obtained concentration-invariant relationships, and the shift factors were obtained by horizontal transfer of the flow curves in the $\lg \eta - \lg \dot{\gamma}$ coordinates.

That the filler particle size can affect the rheological characteristics of composites can be due to several reasons. First, it may be due to the formation of interphases which, as shown earlier, results in increased viscosity. The higher the dispersity of particles the greater the volume of the interphases. However, the decrease of the average particle size does not always result in increased viscosity. The authors of [49] investigated the impact of the particle size of fillers of one type on the viscosity of plasticized PVC. They found that Aerosil A-175 with the mean particle size of 0.02 microns sharply increases the effective viscosity of melts in the entire investigated range of compositions. Addition of 1–2% of white ash with $d = 0.12$ microns, on the contrary, leads to a considerable drop of viscosity. Finally, addition of quartz sand with d of a few microns manifests itself in a monotonous and very insignificant increase of viscosity.

Departing from the known fact that PVC melts have a nodular structure, about 1 micron in size, the authors assumed that white ash particles invaded the inter-nodular spaces, thereby causing the nodules to move apart and the chains between them to be broken; as a result the nodules acquire more freedom to move. As the filler concentration is increased, the contribution to viscosity increase

Table 5. Effect of size of particles $CaCO_3$ and treatment of the filler by stearinic acid on the value of the yield strength τ_y of the molten polycarbonate [162]

Mean size of particles, mm	Specific surface area, m^2/g	Value τ_y for composition with the filler	
		Uncoated	Coated with stearinik acid
0.07	20	38	1.0
0.5	6	9	0.16
3.0	3.6	1.6	0.1
17.0	1.2	0.1	0.1

from the hydrodynamic effects begins to prevail over other mechanisms. If the filler particles are sufficiently smaller than the internodular spaces (which, e.g. is the case for Aerosil), they distribute themselves in the internodular space not loosening, but on the contrary, compacting the matrix structure as a whole. In this case no viscosity decrease effect is observed.

An alternative reason why particle size can affect the polymer melt or solution viscosity consists of the agglomeration of the filler. The higher the dispersity of the filler the higher is its tendency to agglomeration [138–140, 161]. Agglomeration, as will be shown below, affects the properties of filled polymers.

Apart from that, the smaller the particle size, the stronger the structure formed by filler in the melt (that is, the yield strength of polymer τ_y is affected). Table 5 below demonstrates how the yield strength of molten polycarbonate composites depends on the size of $CaCO_3$ particles.

The strength parameters of the structural skeleton of the filler will, in turn, have effect on the flow process of the system as a whole.

Finally, there may be a situation where the addition of a filler with a relatively large initial particle size will entail an even more apparent increase of melt viscosity [163, 164]. Fridman and Popov investigated the effect of foamed perlite particle size on the viscosity of high density polyethylene melt. Three perlite samples were used: small disintegrated perlite (SDP), large disintegrated perlite (LDP), filter perlite (FP). SDP and LDP were prepared by crushing foamed perlite in a disintegrator with four-row pin rotors rotating at 1.5×10^4 and 1×10^4 min^{-1}. The filter perlite (polygonals with a large number of pores and projections and sized up to 60 microns; the particle size distribution pattern is characterized by a spread maximum at about 17.5 microns) was the finer fraction of foamed perlite. Foamed perlite is in fact a mineral foam easily disintegrated when processed in machines. SDP was characterized by a narrow particle distribution with a sharp maximum at $\sim$7.5 microns. The LDP had a broader size scatter (up to $\simeq$40 mm) with the maximum at 13 microns.

Rheological experiments have shown that the relative viscosity of compositions filled with the above materials is an exponential function of filler content by volume. The impact of an otherwise constant quantity of filler increases in the series FP – LDP – SDP, i.e., symbatically with the probability of particle comminution in the plasticization process. This effect is most clearly apparent for fillers coated with a polymeric film by the method of polymerization upon a pre-activated surface [164]. Similar results were also reported in [165].

These experimental data may, presumably, be attributed to the fact that the so-called juvenile surfaces are formed as a result of filler breaking. In the review [22] it is noted that juvenile surfaces enhance the filler-polymer interactions even to the latter's grafting to the mineral particle surfaces.

It should be noted that for polymerization-modified perlite the strength parameters of the composition algo go up with the increasing initial particle size. [164]. In some studies it has been shown that the filler modification effect on the mechanical properties of composites is maximum when only a portion of the filler surface is given the polymerophilic properties (cf., e.g. [166–168]). The reason lies in the specifics of the boundary layer formation in the polymer-filler systems and formation of a secondary filler "network". In principle, the "patchy" polymerophilic behavior of the filler in relation to the matrix should also have place in the failing polymerization-modified perlite.

Particle shape of the filler can also have a significant effect on the rheological properties of filled systems [169, 170].

The characteristic viscosity of suspensions (Einstein's coefficient) of ellipsoid particles increases sharply when the ellipsoid aspect ratio increases [169, 171]. Viscosity anomalies are particularly typical of systems filled with anisodiametric particles. Anomalies are observed even in case of dilute suspension of hard particles in a Newton solvent [172, 173]. The reason is that particles tend to become orientated under shear; whereas Brownian motion, in contrast, causes disorientation of particles. In concentrated suspensions, orientation may go hand in hand with gradual destruction of the filler skeleton, the process being more "extended" through the shear rates (stresses) the smaller the rigidity of the fibrous filler [169, 174, 175]. In capillary viscometric testing (just as in extrusion or pressure molding of articles) the orientation of anisodiametric particles occurs near the measuring channel (or molding tool) inlet more than at any other point; in the channel itself, the opposite may even occur i.e. some disorientation of the anisodiametric filler [176].

For fixed straining conditions of a suspension of anisodiametric particles and filler concentration, the system viscosity will depend on filler hardness [177]: the viscosity decreases, tending asymptotically to a certain limit, as Young's modulus of the filler increases. For this reason the viscosity of systems with polymeric fibers is normally higher than that of similar systems with glass fibers [174, 178]. Of course, apart from filler stiffness, the characteristics of the interphases also have some effect. As a rule the viscosity is higher for longer fibers. For one and the same length/diameter (l/d) ratio of the fiber and fixed concentration of the filler, the relative viscosity is higher for thinner fibers than for thicker ones. [179, 180].

For scaly fillers the increase of relative viscosity with filler concentration is not as pronounced as in case of fibrous fillers [177, 178]. Owing to filler orientation, the flow curves for systems with different concentrations of a fibrous and a scaly filler may merge together at high shear rates [181]. In composites with a dispersed filler the decrease of the effective viscosity of the melt with increasing strain rate is much weaker.

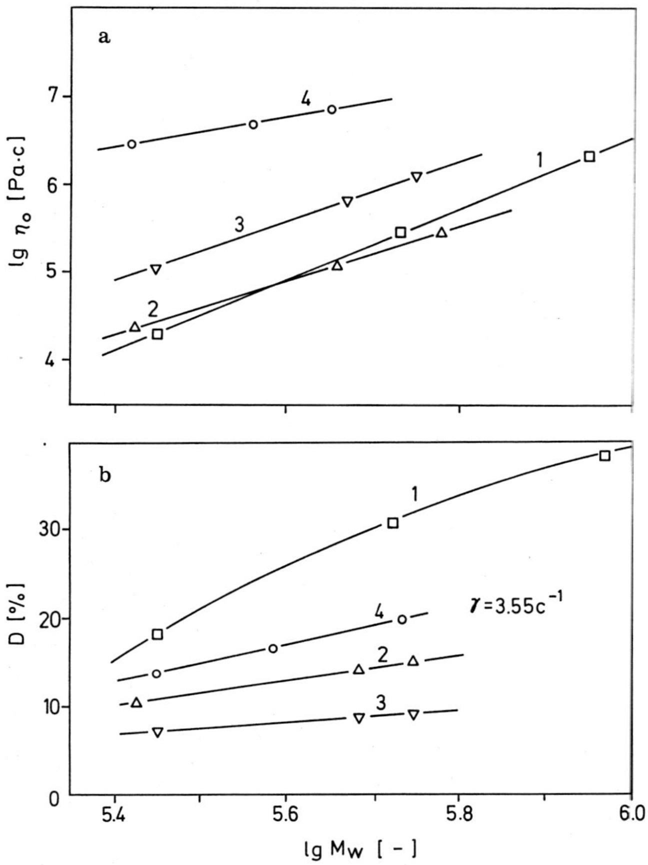

Fig. 2 a, b. Dependence of the maximum Newton viscosity (η_0) (*a*) and swelling ratio of the extrudate (D) (*b*) on the molecular mass of cis-1,4-polyisoprene, unfilled and filled to 33% by mass. *Filler*: 1 — not; 2 — $CaCO_3$ with specific surface areas 2–3 m^2/g; 3 — ash PM–15 with specific surface areas 12–18 m^2/g; 4 — ash PM-100 with specific surface areas 90–100 m^2/g

Note that, apart from the filler particle shape and size, the molecular mass of the base polymer may also have a marked effect on the viscosity of molten composites [182, 183]. The higher the MM of the matrix the less apparent are the variations of relative viscosity with varying filler content. In Fig. 2, borrowed from [183], one can see that the effect of the matrix MM on the viscosity of filled systems decreases with the increasing filler activity. In the quoted reference it has also been shown that the lg η_0 – lg $(MM)_w$ relationships for filled and unfilled systems may intersect. The more branches the polymer has, the stronger is the filler effect on its viscosity. The data for filled high- (HDPE) and low-density polyethylene (LDPE) [164, 182] may serve as an example; the decrease of the molecular mass of LDPE causes a more rapid increase of the relative viscosity of filled systems than in case of HDPE. When the values $(MM)_w$ and $(MM)_w\,(MM)_n^{-1}$ are close, the increased degree of branching results in increase of the relative viscosity of filled system [184].

Filler orientation is not the only consequence of geometric effects in strained systems with an anisodiametric filler. The appearance of normal stresses has been named as another such phenomenon [169, 171, 185] which, like the abnormal viscosity behavior of suspensions, may not be due to the elasticity of system components.

1.7 Filler Effect on Resilience of Polymer Melts

It is assumed that the addition of a hard filler into a polymer matrix should suppress all effects due to the highly elastic properties of the matrix material. The most lucid manifestation of this is seen in the reduced degree of the post-extrusion swelling of the jet (extrudate). This has been observed on many occasions for systems containing both dispersed and fibrous fillers [182, 183, 186–190]. The degree of swelling depends on the elastic energy stored in the materials as it enters the channels and flows along it. It increases the higher the capacity of the system for reversible deformations. Filling reduces the relative volume of the polymer, i.e. of the medium capable of storing the elastic energy of deformation. For ash-filled rubbers [188] it has been shown that if the volume occupied by the filler is taken into account it is sometimes possible to obtain a generalized swelling formula, invariant in respect of the filler concentration. However, the elasticity "suppression" effect cannot be attributed solely to dilution. The filler reduces macromolecular mobility in the interphases and thus increases the melt "stiffness". It is not possible to obtained generalized extrudate swelling curves for systems with fillers characterized by high susceptibility to structurization [188]: in this case the swelling suppression effect is much stronger.

Fibrous fillers dramatically reduce swelling [189], the effect being the more important the greater the relative length of fiber $l \cdot d^{-1}$ [187]. All other conditions being identical, for systems with scaly fillers, the swelling is smaller than for systems with dispersed fillers [177]. The effect of filler particle shape (the difference between an anisodiametric and dispersed filler) also manifests itself in the greater susceptibility of anisodiametric particles, compared to spherical ones, for structurization in the melt and in the easy orientation of anisodiametric particles while disorientation is hindered [181].

In [191] it has been shown that the degree of swelling of the melt $(d \cdot D^{-1})^2$ may be determined from:

$$\alpha = (d \cdot D^{-1})^2 = \beta \tau^A, \tag{29}$$

where d is the diameter; D is extrudate diameter; τ is shear stress on the channel wall; β and A are empirical coefficients.

The value of A decreases linearly with increasing filler content; for a fixed filler content it increases from a filler with the spherical shape of particle to scales and further to fibers. For β the observed behavior was reversed.

Growson and Folkes [187] suggested the formula:

$$\alpha^{-1} = (1 + d_f l_f^{-1})\{1 + [\sigma_{11}(2\tau)^{-1}]^2\}^{1/3} \tag{30}$$

where d_f and l_f are fiber diameter and length respectively; σ_{11} is the first difference of the normal stresses.

With increasing MM the swelling increases faster in systems with a more active filler and is also somewhat greater by an absolute value than in systems with an inactive filler [183].

The capacity of polymer for highly-elastic (reversible) deformations also controls, in large part, the pressure loss effect at the inlet of the forming or measuring instrument. The inlet zone length (l) is related with the highly-elastic reversible deformation S_R by the formula [192]:

$$l = n + S_R \cdot 2^{-1}, \tag{31}$$

where n is the Kuetta correction.

As S_R decreases, l must be decreased too (and thereby also the inlet pressure loss/total pressure ratio is decreased). This is what is really observed when dispersed fillers are added to polymer [182, 190, 193, 194]. The rubber phase in heat resistant polystyrene behaves much like a dispersed filler: it also diminishes the inlet correction [195]. For polystyrene with different fillers the following relationship was found to be valid [196]:

$$[P_{end}(\dot{\gamma})]_c\,[P_{end}(\dot{\gamma})]_m^{-1} = [\tau_w(\dot{\gamma})]_c\,[\tau_w(\dot{\gamma})]_m^{-1}\,(1 - V_f) \tag{32}$$

here, the subscripts m and c relate to the matrix and composite respectively; P_{end} is pressure drop at inlet; τ_w is shear stress on the channel wall; $\dot{\gamma}$ is the effective shear rate. In the quoted reference it was noted that, generally, the pressure drop at the inlet depends on the total pressure, but the filling does not affect the piezo-coefficient of the inlet pressure loss.

Fibrous fillers have often been reported to increase considerably l in comparison with unfilled systems [171, 176, 189, 197, 198]. In this case, however, the inlet loss is due not to the highly elastic properties of the melt but to other reasons, such as pushing of the binder through the "package" (plug) of filler in the inlet zone. Pushing of the filler "package" forth into the channel, etc.

The first difference of normal stresses (σ_{11}) may serve as an indirect index of the highly elastic properties of polymeric systems [199]. C. D. Han [200] related σ_{11} with the residual pressure at outlet P_{exit}. Han, who observed its reduction in polypropylene filled with calcium carbonate [201], concluded that filling decreases the normal stresses. Note that addition of fibrous fillers, vice versa, somewhat increases P_{exit} [180].

Fillers with particles nearly spherical in shape usually dimish σ_{11} under a fixed shear stress [178, 181, 182, 202] in comparison with the pure polymer. In contrast, scales tend to step up the first difference of normal stresses [181]. Fibrous fillers have little effect on σ_{11} value [181], except in a few cases where the tendency for its increase was reported [178, 180], the effect being the more apparent the higher the relative length of fiber, the flexible polymeric fibers increase σ_{11} much more than stiff glass and carbon fibers.

The authors of [203–205] proposed a theory according to which the normal stresses of the matrix and filler may differ only under one condition i.e. the filler content by volume is above some critical value – when its concentration is sufficient to generate the so-called secondary network. In accordance with Privalko' and Lipatov's classification [102], this concentration corresponds to the lower boundary of the high-filled class of composites.

Important for polymer processing is the fact that when the concentration of a hard filler is increased in the composite, the unsteady flow (in the sense of large-scale distortions) of the extrudate occurs at higher shear rates (stresses) than in the case of the base polymer [200, 201, 206]. Moreover, the whirling of the melt flow is even suppressed by small additions of filler [207].

On the other hand, it has been found [208, 209] that for compositions comprising a linear polymer and a certain quantity of a filler one observes a virtually jump-like increase of the volumetric flow rate within a very narrow range of shear stresses, or the so-called flow separation effect. Depending on the size of the measuring capillary, temperature and filler concentration, this effect manifests itself in different forms. When the filler concentration is increased, the relative length of the channel or the material volume are decreased, the flow curves with the typical hysteresis loop change to broken curves. The effective critical stress τ_{cr} (or critical volumetric flow rate Q_{cr}) at which the "separation" occurs reduces as the filler concentration in the system is increased, but the reduction is insignificant. An early separation of a flow of ash-filled rubber was observed in [193] but there has been no study of the influence of different factors on this behavior. A separated flow has also been reported for filled PVC [210]. Note that at stresses a long way below the critical, the filler suppresses the melt tendency to slip along capillary walls [164, 211].

1.8 Effect of Filler Structurization and Agglomeration on Filled Polymer Properties

In a polymeric matrix the filler, even if present in minimum quantities) is always more or less agglomerated [212]. Microscopic studies have confirmed the existence of two types of primary structures in composites [213], i.e., filler aggregates with particles bound together firmly enough, and agglomerates – systems of weakly interrelated aggregates.

As established in [214], in roller-mixed PVC-based composites one observes a much better uniformity of filler distribution over the matrix volume; the relative viscosity is thereby considerably reduced. That is, for a fixed filler content, the viscosity of a system with agglomerates is always higher than that of the "well-dispersed" sample.

Agglomeration also has a strong impact on the stress/strain characteristics of composites. This may be exemplified by the data of Table 6 borrowed from [215].

Table 6. Comparison of ultrasonic control and mechanical testing data for rubber compositions[a]

Characteristic[b]	Rubber formulation		
	SKIZ + SKE 30 ARKP (50:50) + 70 p.b.m. quartz sand PM-75	SKE-30 ARKP + 60 p.b.m. quartz sand PM-15 + 80 p.b.m. chalk	SKE-3C + 35 p.b.m. chalk + 35 p.b.m. kaolin
A_{50}, mV	17.0/37.0	2.7/5.0	7.3/7.5
A_{25}, mV	35.0/52.0	9.0/14.0	16.0/17.0
α, cm^{-1}	29.0/13.5	48.0/41.0	29.4/30.0
σ_{tens}, MPa	8.5/14.2	2.9/4.3	7.4/7.5
ε_{tens}, %	350/400	220/350	420/440

[a] Denominator gives the figure obtained in a rubber mixer; denominator — for additionally rolled mixtures.
[b] A_{50} and A_{25} are signal amplitudes with the 50 and 25 mm base respectively; α is ultrasound absorption factor; σ_{tens} and ε_{tens} are the tensile strength and relative elongation respectively.
p.b.m. parts by mass.

The table data show that the stress/strain properties of compositions are improved by additional dispersion (mixing). Ultrasonic analysis is sufficiently reliable and informative as a means of mixing quality assessment. The very small change of the characteristics for filled compositions (chalk + kaolin) can be due to the fact that these fillers are readily distributed in the matrix as they are.

Nielsen [216] believes an agglomerate to be a plurality of contacting primary particles. Assuming that agglomerates behave like hard particles below a certain stress threshold, the author expects that high stresses should cause the particles constituting an agglomerate to move relative to each other.

Nielsen proposed a formula [216] for estimating the elasticity modulus of a filled polymer:

$$E_c \cdot E_m^{-1} = (1 + ABV_f)(1 - B\varphi V_f), \qquad (33)$$

where

$$A = K_1 - 1, \qquad B = E_m,$$

$$\psi = 1 + [(1 - V_{max})(Vf^2)^{-1}] V_f$$

K_1 is generalized Einstein's coefficient; ψ is reduced concentration dependent on the maximum possible filling ratio V_{max} in a most closely packed system; B is a constant, about equal to 1.

Agglomeration augments the modulus by increasing K_1 and decreasing V_{max}. For cubic packing it has been found, for instance, that agglomeration may cause K_1 to vary from 2.5 to 4.77 and V_{max} from 0.524 to 0.275.

Agglomerates that are poorly wetted by the polymer contain air in their voids and have a lower apparent volumetric content of solids compared to well-wetted particles, the reason being that in the first case the entire matrix can take part in the deformation process, whereas in case of good wetting conditions a portion of the matrix is "caught" in the agglomerate and does not become involved in deformation.

According to Nielsen [217] the elongation at rupture of a filled polymeric composition may be approximately described by the following expression

$$\varepsilon_c \simeq \varepsilon_m(1 - V_f)^{1/3} . \tag{34}$$

Depending on the nature of the polymer-filler interaction and the fracture surface status (smooth or rough), Eq. (34) predicts either a rather smooth variation of the elongation with increasing filler concentration or a sharp drop at some small filler content.

For thermoplastics, polyolefins for example, in contrast, the sharp drop of the relative elongation has been noted at medium filler concentrations [218, 219]. For calcite-filled HDPE, at V_f of up to 0.15 the specimens are deformed much like unfilled polymer [219], the value of Σ_c varying insignificantly in this concentration range and the deformation proceeding in the "bottleneck" manner. At $V_f > 0.3$ the yield point cannot be attained, the failure is brittle and Σ_c becomes almost independent of the filling ratio. In-between, the elongation gradually diminishes, the yield point is apparent, but the specimens are no longer deformed via the "bottleneck" stage. In this intermediate region the failure entails formation of internal pores (a special kind of cavitation) which manifests itself in specimen whitening.

The three deformation regions are also apparent on the strength versus concentration relationships. The most dramatic drop of the yield point was observed at small filler concentrations (up to 0.15). On further filling the characteristic remained almost unchanged.

In [220] it was shown, for spheres with a radius tending to zero, that there must be a critical value of $B_{cr} = 4 \cdot 3^{-1}\pi a^3 N$ in the system which is conductive to an infinite cluster of bonded sites (here a is site spacing; N is volumetric concentration).

To determine the geometric properties of systems with spheres, the authors of [221, 222] carried out a Monte-Carlo simulation of volume filling. In [221] they showed that for $ar^{-1} = 1.0 - 0.8$, the critical filler concentration was 0.1 to 0.13. Calculation has shown that at small concentrations the particles belong, in large part, to the small-sized clusters, whereas at higher concentrations the size distribution of the clusters broadens. Further increase of filler content sharply increases the probability that particles will agglomerate into one large cluster. Figure 3 shows the filler concentration dependences of the relative elongation at rupture [219] and probable cluster sizes at $ar^{-1} = 1$ [221]. The sharp drop of the stress strain parameters occurs precisely at the point of transition to one integral cluster. The formation of a range of different clusters in the transition region brings about an increased scatter of the limiting strength parameters.

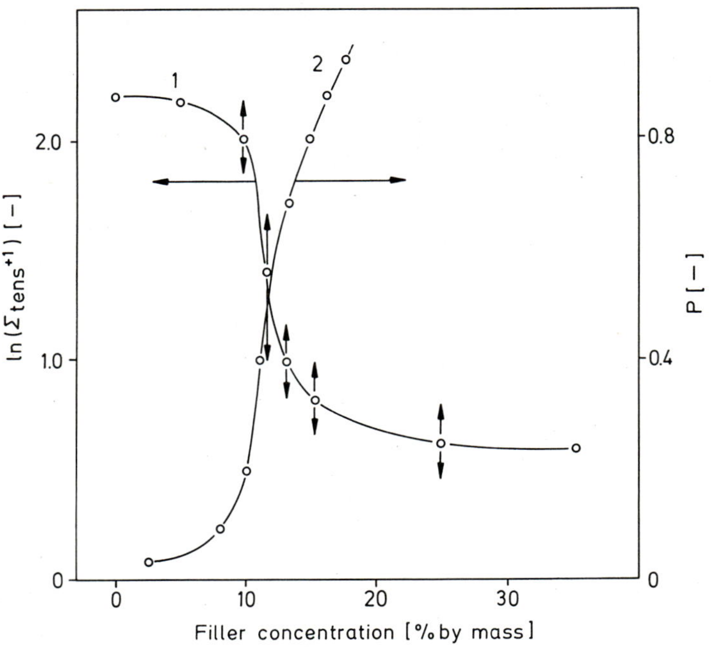

Fig. 3. Concentration dependences of deformation at rupture ln (Σ_{tens} + 1) of the HDPE + glass balls system (**1**) and maximum cluster size P (**2**) for ar^{-1} = 1.0 composites [219, 221]

Analysis of the relationships between the moduli and bond strength between particles [222] has shown that for $V_f = 0.1 - 0.15$ the concentration dependence of the modulus corresponds to the lower curve in the Hashin-Shtrikman equation [223] (hard inclusion in elastic matrix), and for $V_f = 0.34$ to the upper boundary (elastic inclusion in a hard matrix). The 0.1 to 0.34 range is the phase inversion region.

Heat treatment at 400 °C of pure polymer specimens and composites containing 0.16 and 0.34 of calcite has shown [221] that the base polymer was sublimated without residue; the specimen with $V_f = 0.16$ left a powder, and the specimen with $V_f = 0.34$ a brittle skeleton. Electron microscopy confirms that the latter consists of large crystalline inclusions interconnected by systems of small particles.

The elastic properties of viscoelastic filled polymeric systems are the first to be affected at critical filler concentrations, due to formation of a spatial structural network [224]. The strength of this network, or skeleton, depends on the force of cohesion and the number of contacts between the filler particles, either direct or via matrix layers [169]. As is well-known, the flow curves of such systems may have three distinct regions [225, 226]: the flow (creep) region structurally undestroyed material, the yield point region (τ_y), and the flow of destructured composition. The yield point is independent of temperature, increases with increasing filler concentration and decreasing particle diameter [162, 228–230]. In principle, the flow curves of filled polymers are the same for all systems and has little dependence on the nature and viscosity of the matrix [229, 231]. Transition across the yield point entails a sharp (sometimes several orders in magnitude) drop of the effective viscosity which may result in the brittle failure of the filler skeleton. Where the interparticle bonds are not very secure, or the filler itself possesses viscoelastic characteristics, the failure process extends in time and depends on the applied stress [232]. In a low-molecular matrix the transition across the yield point is a rather sudden process, whereas in a high-molecular matrix it is "masked" by the non-Newtonian behavior of the base polymer [229, 232].

The strength of the filler skeleton may be characterized by the complex dynamic modulus measured at low frequencies [24]. The authors note that when $c > c_{cr}$ the yield point should be viewed as a sum of two components:

$$[G^*]_{c > c_{cr}} = [G^*]_{c \leqq c_{cr}} + A(c)\,, \tag{35}$$

where the first term is the complex modulus value at critical filler concentration; $A_{(c)}$ is the term due to the contribution of the secondary agglomerates. –

In the first approximation $A_{(c)}$ should be dependent on the molecular characteristics of the matrix and matrix-filler interaction.

It should be noted that the ability of the filler to form the structural network is largely responsible for its reinforcing action [28, 33]. These aspects of the problem have been discussed at length in a number of monographs (e.g. [7, 53, 54]) and we do not think it proper to dwell on them here. On the other hand the structurization in melt increases the viscosity of the material and hampers its processability.

To improve the moldability of polymeric composites technologists have proposed a number of methods, such as molding under a complex shear [223]. In melts, the complex shear may be effected by rotation, longitudinal or transverse (relative to the extrusion axis) vibrations of the working elements of the molting instrument. The simplest example of complex shear is the spiral flow arising when the material is extruded through a ring-shaped orifice under a constant pressure difference and rotation of the cylinders (or one of them, for example, the nozzle or mandrel of the extrusion die).

For filled thermoplastics (30–40% by mass of chalk, ash or asbestos), complex shear may, as reported in [235], provide an increase of apparatus productivity by 40–80%, or if the flow rate is to be constant, the pressure in the molding instrument may be reduced by at least 20–30%. It is to be noted that while some extra power is required to create the complex shear conditions, the total power consumption of the apparatus as a whole may be reduced, on the power per unit of product basis, due to the high extrusion rates [233].

For all that, various modification methods remain the mainline trend in PCM property control.

1.9 Effect of Adhesion on PCM Properties. Property Control by Modification

In an overwhelming majority of investigations into the effect of fillers on the mechanical properties of polymeric materials (cf., e.g. [234]) it was noted that the different behavior of systems is largely due to the adhesion between the filler and the polymer.

This may be exemplified by the results of [235] in which the properties of PE-based compositions with $CaCO_3$ and kaolin with particles of about the same size were studied. Kaolin had a much greater adhesion to PE than calcium carbonate, as $CaCO_3$ manifested by the presence of polymer bridges on kaolin particles after specimen failure and their absence on calcium carbonate particles. The systems behaved differently. For kaolin-filled compositions the yield point (σ_y) remained invariable up to 40% by mass of filler. However, at higher percentages, the composites failed at elongations (ε) much smaller than the elongation at σ_y. The relative elongation did not change until 30% and then dropped very suddenly. For $CaCO_3$, ε varied only slightly in the same concentration range, but as filler content increased the σ_y decreased. The impact strength for kaolin filler was much more concentration dependent than for calcium carbonate.

From the models worked out in [236, 237] it ensues that for a fixed filler content the composite moduli must be greater in the case of absolute adhesion than where there is practically no adhesion at all. Yet, in a number of works it has been noted that, quite to the contrary, the modulus of the composite is practically independent of adhesion [238–241].

Lee [242] studied the dependence of the physico-mechanical properties of Wollastonite-filled polychloroprene rubber on the type of agent used to pre-treat the filler. The composition contained 26.9 part (weight) of the filler per 100 parts (weight) of the rubber (compositions CR-1100, CR-174, CR-151). The finishing agents were γ-aminopropyl triethoxysilane (CR-1100 and CR-174) and vinyl triethoxysilane (CR-151). The mechanical properties of the compositions are listed in Table 7 below. The author proposed an empirical equation to relate the modulus with the equilibrium work of adhesion in the following form:

$$\ln E = -4.80 + 1.267 \ln W_A . \tag{35}$$

Table 7. Effect of Wollastonite treatment method on the mechanical properties of filled chloroprene [242]

Property	Composition		
	CR-1100	CR-174	CR-151
Filler/polymer surface energy ratio	1.14	0.87	0.76
Equilibrium work of adhesion (W_a), MPa · m	70.80	61.88	57.58
Young's modulus	1.83	1.56	1.35
Tear strength, kN/m	5.82	5.06	7.06
Tensile strength, mPa	12.3	13.1	13.3
Relative elongation at rupture, %	641	656	703
Effective viscosity logarithm at shear rate of 1 s^{-1}, kPa s	5.82	5.06	4.06

Table 8. Heats of interaction and physico-mechanical properties of PVC-based filled plasticized compositions[a] [100]

Filler	$\Delta H \times 10^{-3}$ J/kg of filler	D MPa	E MPa	σ_0 MPa	ε_0 MPa	σ_{tens} MPa
A-175	−6.0	34.3	4.1	9.8	85	12.3
Butasil	−2.8	24.0	—	13.8	100	16.3
AM-2	+2.2	22.1	2.6	17.9	130	19.1

[a] The stress σ_0 and elongation of separation ε_0 were determined for composites with 16% by mass of the filler; other characteristics with 22% by mass of filler.

Similar results were also reported in [100]. As noted earlier, the authors measured the strength of the filler-matrix interaction by the magnitude of the thermal effect: the more negative the enthalpy, the greater the interaction. Pure A-175 Aerosil and Aerosils modified with aminophenyl methylene diethoxysilane (AM-2) and butyl alcohol ("Butasil"). The base polymer was PVC plasticized with dioctyl sebacinate. The results are presented in Table 8 below.

As in the previous case, the increased filler-matrix interaction bring about an increase in both dynamic (D) and equilibrium (E) high elasticity moduli.

Obviously, the discrepancy between the experimental data [238–241] and predictions of the theory [236, 237] can be attributed to the difference of the coefficients of thermal expansion. The polymer exerts pressure on the filler, thereby masking the effect of the strength of adhesion on the modulus. The pressure on the filler may be sufficiently high. In [243] it was found, for example, that in PP, quartz particles experienced a compression force of about 100 MPa; after cold drawing of the composite the force reduces to 50 MPa in the direction of drawing but at the same time increases to 300 MPa in the perpendicular direction.

In a series of reviews [244–246] the models proposed for the assessment of the effect of fillers on the complex of PCM properties are discussed. Analysis of the models shows that, for a fixed filler content, the strength must be higher in compositions with fillers featuring the absolute adhesion to the matrix than in systems with little or no adhesion. The relative elongation and specific impact strength must, on the contrary, go up with the increasing adhesion.

According to [116], the relationship between tensile strength and the strength of the adhesion bond between the filler and polymer may be estimated as:

$$\sigma_c = \sigma_A V_f^{2/3} + \sigma_m (1 - V_f^{2/3}) . \qquad (36)$$

Even though Eq. (36) was derived on the basis of a simplest model and, as the authors themselves admit, cannot be expected to provide a high degree of agreement with the experiment, it can well be used for obtaining rough estimates of the matrix-filler interaction.

Today, we have well-tried methods of experimental assessment of the shear strength at the fiber-binder interphase [247]. Generalization of a wealth of experimental data [247, 248] shows that for inorganic fiber-filled PCM an increase of strength at the interphase entails a symbatic increase of the overall strength of the composite material. Reduction of adhesion strength has a most significant effect on the shear strength indices. Arranged in the order of increasing sensitivity to deterioration of the polymer-fiber bond, the types of mechanical testing are [247]: tension < bending, < compression < shear. At the same time, the strength of organic fiber reinforced plastics is more often than not independent of the force of cohesion between the components [247, 248], remaining practically invariable under any kind of stressing. The authors of the references quoted above relate this behavior with the fiber structure and material behavior in failure. The individual fibers used in the experiments consisted of a plurality of highly orientated and very strong fibrils weakly associated with one another. The strength of such a composition is not much greater, even if loaded along the fibers, by the strength of the fibers themselves, since the failure process is limited by the strength of the bond between the fibrils themselves.

Impact strength also increased if the adhesion between the polymer and fiber is increased [240, 249]. The most promising method of modification of fiber-filled compositions is by pre-treating the fibers or adding to the matrix of specific depressants or modifiers with the aim of creating a chemical bond at the interphase. This improves the composition service lifetime, strength and thermal stability [250].

In investigations of the failure of fiber compositions (PETP – short glass fibers) [251] it was found that the main process responsible for composite failure under load is the rupture at the matrix-fiber interface. The author of [251] observed formation of microvoids in loaded samples, both at the interphases and in the bulk. The microvoids, or cavities) grow in size and become interconnected by microcracks, and this results in fiber separation from the binder. However, when the matrix-fiber bond is strong enough, the cavities appear mostly in the bulk of matrix, the failure of the specimen does not "over-power" cohesion and traces of polymer remain on the fibers.

In [252–254] the variation of specimen volume due to deformation was determined for a number of thermoplastics filled with differently pre-treated fibers using the formula $\Delta V \cdot V_0^{-1}$ (here, ΔV is volume increment, V_0 is initial volume); this was done by measuring the longitudinal and transverse deformations. It was found that some degree of loosening was typical not only of the matrix but the composite too. The authors considered the degree of loosening as a function of temperature and found that the loosening was maximum in composites precisely at temperatures at which the loosening of the matrix was minimum, that is, in the temperature ranges where the deformability of the composite is, apparently, due to the separation of fiber from the matrix. The loosening was lowest in composites where the fibers were caused to be chemically bonded to the matrix.

The failure of systems with dispersed fillers (exemplified by polystyrene plus glass spheres with different treatment) was studied by subjecting specimens to deformation in the microscope field [255, 256]. Where adhesion was good the cracks were observed to be formed near the glass sphere pole, in regions corresponding to the maximum deformation, where adhesion was poor, anywhere between the pole and the equator. It was discovered that microcracks began to be formed only after the separation of the matrix from the filler which gradually propagated from the pole towards the equator. After a separation of $\simeq 60°$ has developed in a zone, cracks begin to appear at the boundary, which may be partially healed at rest.

In loading experiments the separation of the matrix from the filler is one of the reasons responsible for the deviation of the stretching diagram $\sigma_{tens} = f(\varepsilon_{tens})$ from linearity. In case of poor adhesion, the stresses which cause such deviation from linearity are lower than in case of good adhesion [240]. For good adhesion, the value of ε gradually decreases with increasing filler concentration; if adhesion is poor, it remains invariant up to a certain concentration and then drops very suddenly.

Poor mechanical properties of rubber products may also be due to matrix separation [257]. Just as in other systems, separation gives rise to cavities and initiates failure. These processes prevail in systems with poor adhesion and become more probable with the increasing filler modulus.

It is quite a long time ago, now that Tshoegl [258] showed that the strength of filled systems could be greatly improved if the system were subjected to a hydrostatic pressure, whereby matrix separation is prevented even in systems with zero polymer-filler adhesion.

The authors of a very recent work [259] simulated the filler effect by applying an effective tensile stress σ_Z in the direction perpendicular to the direction of the tensile stress σ_T. To determine σ_Z they relied on the condition of equality of the work of external forces which cause deformation of a unit volume of the composite and matrix subjected to the above-mentioned stresses (asymmetric tri-axial stretching). They came up with a formula of the form:

$$E_c\varepsilon_c^2 = E_m\varepsilon_m^2(1 - V_f)[1 + Q_m\sigma_Z(\sigma_1 - 2V_m\sigma_Z)^{-1}], \tag{37}$$

where ε_c is composite deformation; $\varepsilon_m = \varepsilon_c(1 - V_f)^{-1}$ is matrix deformation; Q_m is Poisson's ratio of the matrix. Hence,

$$\sigma_Z = \sigma_1(2Q_m)^{-1}\{1 - E_m[E_c(1 - V_f)]^{-1}\}. \tag{38}$$

In other words, σ_Z can be determined from the tensile stress/strain diagrams of the filler and the matrix.

In [259] it has also been noted that the stressed state arising in the "bottleneck" portion of the stretched cylindrical specimen is the same as in the model and that specimens with the "bottleneck" portion of specific geometry may be expected to fail in the brittle manner. The authors calculated that a hydrostatic pressure of 8–12 MPa (depending on straining rate) should be enough to compensate for the tensile stress σ_Z in a PE + 40% by mass of chalk system. They found that such hydrostatic pressure sharply increased the elongation at rupture; the elasticity modulus and yield point also rise. As supplement to Tschoegl's method they demonstrated that ε could, alternatively, be considerably increased by rolling filled specimens; the maximum ε were attained with rolling ratios k ($k = b_0b^{-1}$, where b_0 is the initial specimen thickness) of between 2.3–2.5.

Adhesion is usually controlled by means of various finishing agents. Mikhalsky noted in [260] that reactions between such agents and thermoplastics are hindered for a number of reasons, one reason being that the chemical structure of the polymer is formed before the treated filler is added. In the majority of cases thermoplastics do not contain reactive groups, if perhaps only at the ends of macromolecules where they enjoy little mobility. The probability of contact between the reactive groups of the agent and the plastic.

Further means used to enhance the effect of finishing agents include [260]: activation of the chemical reaction between the agent and the thermoplastic with the help of peroxides; grafting to polymer macromolecules of groups known to be active in relation to the finishing agent; creating an interlayer between the treated filler and the thermoplastic matrix that would feature the "double" affinity to both the finishing agent and the matrix.

Finishing of the filler surfaces may also greatly affect the system viscosity. For mica-filled PP [31] and various thermoplastics filled with calcium carbonate [202, 261] it was shown that the relative viscosity of filled systems was lower than that of systems which contained equivalent quantitied of unfinished filler. Note that in contrast to viscosity in shear, the viscosity in stretching is higher for systems with treated filler [202].

Organotitanates form regular adsorbed layers on the filler surfaces. This assures a high degree of dispersibility of the solid particles of the filler, removal of moisture and air from the surfaces, a material improvement of the rheological properties of filled compositions. Also, it is possible to use much greater percentages of cheap filler [244, 245, 262].

Using water repelling agents it is possible to either considerably reduce the viscosity of compositions with a predetermined filler content, or increase the admissible filler concentration for given viscosity [262]. In a filled system (e.g. PP + chalk) the treatment of the filler to enhance its polymerophilic properties promotes breakdown of agglomerates [263].

In the case of PE + $CaCO_3$, the critical volumetric concentration of filler up to which no reduction of the impact strength of the composite occurs, increases markedly if 1–3% organotitanate are stuck on the filler surface [244, 245]. It has been found possible to considerably improve the impact strength of polyolefins by treating the $CaCO_3$ particles with ethylene glycol or ethylene oxide oligomer [264, 265]; this however is accompanied by reduction of the modulus and tensile strength, but the relative elongation is increased. The authors of [244, 245] attribute the increased impact strength to formation of a thin elastic layer of sufficiently high molecular mass. The latter promotes rapid redistribution of loads and releases the internal stresses. Besides, such layers provide conditions for a more intensive dissipation of heat in a specimen, and thereby the development of microcracks promoted by the released heat is hindered.

Yet, some authors reported an increase, instead of reduction of viscosity after finishing treatment of filler [266, 267]. Plueddeman in [268] noted that increase or decrease of melt viscosity with finishing agents depends on what is the resultant reaction of the filler, agent and matrix — neutral, acid or basic. The viscosity may be optimized by using the right formulation. For example, agents with basic properties should be used for acidic filler only if the polymeric matrix is substantially neutral. By choosing the filler-polymer system on the above principle it is also possible to minimizing the ingredient mixing work in preparation of composites [269]. Some investigators believe [270] that finishing agents improve processability of compositions and reduce the wear of molds, whereas others (cf., e.g. [271]) insist that the increased adhesion of polymer to filler following the finishing treatment may greatly increase the abrasive characteristics of the composition. If the unfinished mineral particles is not secured in the matrix it may with relative ease be peeled off the contacting metal wall of the mold, or even roll like the ball of a ball-bearing. Finishing causes the particle to be secured in place in the matrix, its edges (this is especially true for particles of irregular shape with sharp edges) fixed in the matrix like a razor blade in a holder and capable of cutting (indenting) the metal walls of the mold. This mechanistic representation is, of course, a gross oversimplification of the real situation, more appropriate for a solid composite than a melt, but nevertheless is, in our opinion, not totally inadequate.

By now, the fluctuation mechanism of the buildup of broken interatomic bonds and generation of incipient cracks has received more evidence in its favor than any other mechanism, and the so-called dilaton model has been developed on its basis [272]. The low density regions in the polymer-filler interphase feature a greater phonon capturing capacity [273] which results in an increased local phonon temperature of dilaton and its consequent thermal expansion. If there is a constant energy flux into the system, such as that generated in a strained specimen, the dilaton size becomes critically large which leads to its ultimate breakdown and incipience of a microcrack. The dilaton development kinetics depend largely on the relative rates of energy supply and dilaton dissolution.

Earlier in [274] it was proposed that the increase of polymer surface by the introduced filler plays the decisive role in the formation of regions characterized

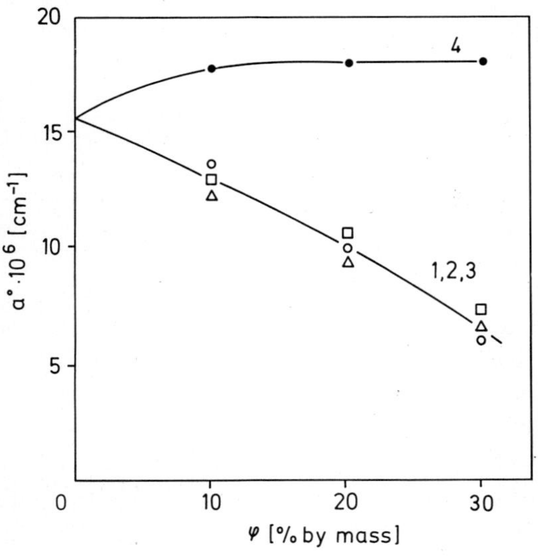

Fig. 4. Dependence on filler concentration (φ) of the coefficient of absorption a^0 at the 730 cm^{-1} peak for volatile ash fractions. *Filler*: 1 — untreatment; 2 — treatment calcium stearate; 3 — treatment PPES; 4 — treatment atactic PP

by a high phonon capturing ability, since the molecular interaction of polymer is lower on the filler surface than in the bulk. It has also been hypothesized [275] that the dilaton dissolution rate can be purposedly controlled, and all that is needed for that is only to provide conditions for the effective dissipation of energy at the interphase by, for example, adding a noncrystallizable agent which will enhance both the conformational heat conductivity of the amorphous boundary layer and the intermolecular interaction of polymer on the filler surface. Using HDPE compositions filled with volatile ash fractions the authors of [275] studied the effect of coating this kind of filler with different materials on the physico-mechanical properties of the composite. The finishers were: calcium stearate, poly(phenyl ethoxysilane), amorphous (atactic) PP. The samples were extruded and then pressed into films. Figures 4 and 5 show limiting cold drawing ratios and absorption coefficient at the peak of 730 cm^{-1} which characterizes the intermolecular interaction as functions of filler concentration for compositions with different additives. One can see that the deformability and intermolecular interaction vary symbatically; treatment with atactic PP gave a material which could be deformed via the "bottleneck" stage and could be drawn with high ratios. The authors believe that in this case the matrix did not suffer the substantial hindrances it normally experiences when there is a strong adhesion bond between the polymer and filler.

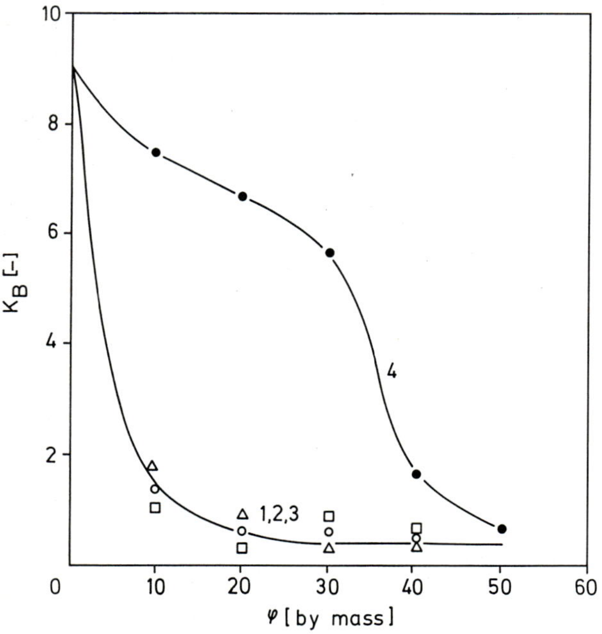

Fig. 5. Dependence of the limiting cold drawing ratio (K_B) on filler concentration (φ) for volatile ash fractions. *Filler*: 1 — untreatment; 2 — treatment calcium stearate; 3 — treatment PPES; 4 — treatment atactic PP

Table 9. Effect of silane treatment on the mechanical properties of kaolin-filled polyethylene compositions [276]

Silane	Silane concentration		σ_{tens} MPa	σ_{bend} MPa	Impact strength, kJ/m^2
	g/100 g of kaolin	k average, %			
PE + kaolin	–	–	19.4	37.7	1.79 ± 0.15
I	1.76	0.5	20.3	38.8	1.78 ± 0.19
I	7.0	2.0	23.5	42.4	1.73 ± 0.03
II	2.75	0.6	24.2	49.5	2.12 ± 0.16

In [276] the properties of kaolin-filled polyethylenes were investigated. The filler was pre-treated with silanes of two kinds, vinyl triethoxysilane (VTES) (I) and gamma-methacryloxypropyl trimethoxysilane (II). The results are presented in Table. 9.

Note that in case of VTES (I) there is a chemical interaction via the vinyl group between silane and the filler, which results in a sufficiently "rigid" bond between the matrix and filler. The agent (II) undergoes homopolymerization so that an elastic sublayer ("shell") is formed around each filler particle, the tensile and impact strength of the composition increase as a result.

Thus a strong bond is not always desirable. We can see this from Table 7 and 8. The authors of [100] interpreted their experimental data as follows: the rigidity of specimens increases with increasing PVC-filler interaction; as a result the rate of relaxation of stresses arising at interphases in the course of deformation decreases. The overstressed states at the interphases may, in the authors' opinion, promote separation of the polymer from the filler surface. That is, it is more desirable that the matrix-filler bond is not rigid but labile.

Rigid bonds between the filler and the matrix cause microstresses to appear in the binder, which are the higher the more rigid the polymer and the greater the difference between it and the filler as regards the coefficient of thermal expansion [277]. To bring down the local stresses it was suggested in [278] to create an elastic interphase, so that plastic materials with the desired static and dynamic strength parameters can be produced by proper adjustment of the thickness and other characteristics of the elastic interlayer for each type of composition. According to [279] the elastic interphase can be used to the best advantage when its modulus is about 0.1 of the matrix modulus. It has been proposed to apply coat on just a certain portion of the filler, for example, leave uncoated segments on the fibers. As an example it can be pointed out that elastic coats considerably upgrade the mechanical characteristics of filled polycarbonates [281]. With layers of organosiloxane block-copolymers of different molecular mass it has been possible to improve considerably the cracking resistance of glass-filled phenol-formaldehyde resins when exposed to liquid nitrogen. The effect was greater when block-copolymers of higher molecular mass were used [282].

2 Properties of Polymerization-Filled Composites and Compositions with Polymerization-Modified Fillers

2.1 Preparation of Polymerization-Filled Composites

This relatively new trend in PCM manufacturing business amounts to creating a polymeric matrix out of the liquid or gaseous phase directly on the filler surface which has previously undergone special conditioning aimed at generating active polymerization sites on it.

The polymerization filling was effected by the ion-coordination mechanism [17–19]. The monomers were ethylene, propylene, allene, α-butylene, butadiene. The fillers were mineral materials such as ash, graphite, silica gel, glass fibers. The ultimate aim of filler conditioning prior to polymerization is to secure, on its surface, metal complex or organometallic catalysts by either physical or chemical methods [17–19].

The physical methods consist of depositing, in one way or another, a layer of transition metal compounds on the surface and pores of the filler, the presence of functional groups on the surface not being necessary. The chemical methods are based [17–19] on causing the transition metal or organo-metallic compounds to react with the reactive groups (e.g. hydroxyls) available on the filler surface. The surface density of hydroxyl groups can be easily controlled by chemical or heat treatment. If no functional groups are available, one of the preparatory stages will be graft a minor quantity (0.1 to 2%) of the monomer with functional groups. Other ways of producing catalytic systems are possible, too, for example, it was shown possible to prepare PE with a filler-catalyst made by applying organic compounds and hydrides of Ti, Cr, Zn onto natural alumosilicates and active in the absence of an organo-aluminum co-catalyst [283]. In [284] it was shown that calcined kaolin exhibited catalytic activity in polyethylene polymerization thanks to the presence of TiO_2.

Titanium, vanadium or chromium oxides activated with chlorine-free organo-aluminum compounds, triethyl- or triisobutyl aluminum, have also been used as catalysts [285].

It should be noted that by immobilizing the metal complex catalysts on carbochain polymers it is possible to prepare polymer-polymer compositions using PVC, PMMA, poly(vinyl acetate), etc., as the polymeric fillers [286–287].

Recently, a synthesis of PE on the waste of the cellulose and woodworking industry has been reported [288]. Good quality structural and cladding materials can be obtained by pressing these compositions, or special grades of paper and cardboard after appropriate processing.

The filler used in polymerization filling or that to which a polymer is grafted by the radical or ion mechanism is activated either physically or through chemical modification of the solid surface [20–22]. Physical activation can be done by such methods as milling and dispersing in the presence of the monomer, plasma treatment, and exposure to ionizing radiation [20–22]. Chemical activation involves securing on the filler surface of polymerizable groups such as peroxides [289–290].

The monomers may in this case be vinyl chloride, styrene, methyl methacrylate, etc. [20–22].

As shown for the synthesis of PS [291], the monomer may be localized in the vicinity of the filler surface by previously grafting a polymer capable of swelling in the base monomer. Copolymeric latex of polychloroprenemethacrylic acid was added to the aqueous dispersion of chalk. The acid groups reacted with chalk and the latex particles became chemically grafted to chalk. When further portions of styrene were added they were completely absorbed by modified chalk.

2.2 Structure of Polymerization-Filled Composites

As a result of polymerization filling a polymer film is formed around some of the filler particles or agglomerates the structure of which depends on the nature of the monomer, method of activation of polymerization, quantity of the polymer formed in the process and other factors.

It has been found [292] that agglomeration of filler may result from securing the catalyst for carrying out the polymerization by the ion coordination mechanism. When the catalyst is deposited from the liquid phase the mean filler particle size may be 3 to 6 times as great as the initial size depending on the manner in which the catalytic complex was formed. Application of the catalytic from the gas phase does not entail such a considerable agglomeration. The agglomerates have a loose skeletal structure with a large pore size and developed internal surface on which the polymerization actually occurs [292]. Further agglomeration may result from the formation of the polymer coat, whereby the final particle size may be much greater than the initial size [293]. Despite agglomeration, practically the entire surfaces of the filler particles gets coated with the polymer [17, 294].

Electron microscopic analysis has allowed us to establish that in polymerization-filled PE with 10–30% by mass of polymer the latter envelops the filler particle with a thin continuous film [295–298]. In composites containing more than 30% by mass of polymer, submolecular structures have been detected [294, 297]. The resultant morphological picture depends both on the polymer film thickness and the type of filler used. For example, for perlite (foamed volcanic glass) the prevalent process was the incipience and growth first of globular and then columnar structures; for kaolin and tufa a small number of spherical structures. In thicker polymeric coats (polymer content of more than 30% by mass) one can clearly see small cracks. In the cracked areas, the monomer diffusion to the catalyst is easy and results in a very uneven film thickness. The authors of [295] assume that in course of polymerization the newly formed layers cause separation of primary layers and their displacement from the surface. As a consequence, different layers may have different cohesion with the filler surface. At concentrations above 50% by mass they observed formation of long "worm"-like secondary particles of polymer which did not contain the filler [296]. Apparently, the appearance of such particles could be due to the partial cracking of the primary coat by virtue of the overstresses produced by polymerization, as proposed in [295]. This is followed by the free growth of "worm"-like and spiral-shaped polymer particles fixed at one end on the catalyst.

Polymerization filling considerably increases the area of contact between the matrix and the filler in comparison with conventional mixing techniques, because the monomer can penetrate the tiniest cracks and pores in the filler [299, 300], whereby the bonds between the components are strengthened. The polymerization of PP on porous filler was accompanied [301] by formation of microfibrils with micromolecules orientated normally in relation to the filler surface. The degree of orientation is high enough and is about that which arises from four-five fold drawing of fiber made of the same PP grade.

An interesting effect has been described in [297, 298]. In experiments involving thermal expansion of polymerization-filled composite material (PFCM) based on PE and containing 80–95% by mass of filler it was found (cf. Fig. 6) that, in contrast to compositions of similar formulation but prepared by mixing powders of the filler and superhigh molecular polyethylene (SHMPE), the $\alpha = \Delta l(l \cdot \Delta T)^{-1}$ relationship of PFCM (here l is the current height of specimen, Δl and ΔT are increments of specimen height and temperature, respectively) contains two typical peaks. The first peak occurs at temperatures between 125 and 135 °C, the second at 180–190 °C. With the decreasing percentage of polymer in PFCM, the height of the first peak decreases and of the second increases. For the standard polymer-filler mixture, there is only one peak at 125–135 °C. By microcalorimetric analysis it was found that the first peak corresponds to the endo- and the second to the exo-effect. The first maximum is due to polymer melting, the nature of the second is not altogether clear except that it must be associated with the fine aspects of the structure and morphology of the polymer on the mineral particle surfaces in PFCM.

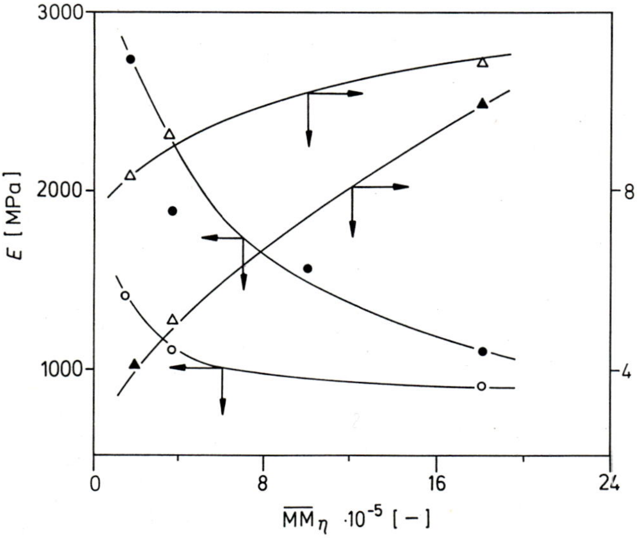

Fig. 6. Variation of elasticity modulus (E) under tension and yield strain (ε_y) of the polymer matrix (I, I′) and polyethylene-based composites polymerization filled with kaolin (2,2′) in function of polymer MM [320]. Kaolin content 30% by mass. The specimens were pressed; 0.3–0.4 mm thick blates; stretching rate $\dot{\varepsilon} = 0.67$ min^{-1}

It was also shown that, given equal filling percentages, the PFCM have a lower coefficient of linear expansion than the composites prepared by conventional mixing. This may be due to the closer packing of macromolecules on the filler surface. The polymer film formed on the filler particle in the process of polymerization filling has a higher density than either unfilled polymer or a composite obtained by conventional mixing [302–304].

For a fixed filling ratio, the degree of crystallinity and mean crystallite size are somewhat higher in PFCM than in mechanical mixtures of similar composition and similar matrix characteristics [299, 300].

In polymer regions of PFCM bordering on the filler one typically observes the formation of crystallites as tape-like structures orientated perpendicularly to the filler surface [295, 305, 306]. In real fact these are transcrystallite structures. A morphology of this kind can appear in a mechanical mixture as well [305, 306], but there the layer thickness will be much smaller [306, 307].

The phenomenon of decrease of the relative dynamic shear modulus of filled systems with increasing temperature is well known [308]. The authors of this work attributed it to agglomeration of filler particles. In contrast, no such relationship was observed for PFCM (in PP-graphite samples, for example) [307] which may be indirect evidence of the high degree of filler dispersion in the matrix. It has been found [309] that PP synthesized on graphite features high electric conductivity which in some cases was several orders of magnitude higher than that of conventionally prepared composites containing the same ingredients. Earlier it had been determined that application of a high pressure in combination with shear strain to metal-filled polymeric systems caused it suddenly to pass into the high-conductive state [310]. In [311] the pressure was measured at which this jump-like drop of electric resistance and increase of conductivity occurred. Compositions with 7% of graphite were used prepared conventionally and by chemical (polymerization) filling. The specimens were cooled under the tempering and annealing conditions. The threshold pressures were found to be 190 and 170 MPa for mechanical mixtures, and 160 and 50 MPa for PFCM (for tempering and annealing, respectively).

On the basis of the above data it has been hypothesized that the conductivity of PFCM is due not to the contact between the filler particles but the current passes across the thin (less than 1–2 microns) polymer interlayers. The conductivity arises when a spontaneous pressure exceeding the threshold value develops in the material. The overstresses apparently arise as a result of PP crystallization in the very narrow gaps between the filler particles [312]. Since crystallization must strongly affect the macromolecular conformation whereas the narrowness of the gap and fixed position of molecules on the filler prevent it, the heat released in the process of crystallization must, in part, be spent to overcome this hindrance, whereby a local high pressure may arise in the gap. This effect is possible only where there are gaps of the size comparable with that of macromolecules. The small gap thickness will also hamper pressure relaxation, since the rate of flow from such a narrow clearance should be negligibly small.

The polymeric coat formed on the substrate by radical and anion polymerization has been observed to be mosaic [313–315]. The polymer was there in the form of globules, or even drops of microscopic dimensions. In [22] it has been suggested that this might be due to the nonequivalence of the active sites which initiate graft polymerization in terms of energy. It has also been assumed that the continuity of coating on the individual particles can only be possible with sufficient degrees of monomer conversion on the surface, when the separate drops of polymer merge together, or by virtue of homopolymerization. Alternative situations are, obviously, also possible. For example, globular structures of the number and size depending on the reaction conditions and time have been observed to result from polymerization of vinyl chloride on the surface of chalk activated with peroxysuccinic or acrylic acid [316]. By removing first the homopolymer and then the filler by extraction with dichlorethane and hydrochloric acid, respectively, it has been possible to establish that the grafted polymers are formed as a continuous film.

The authors of [317] investigated compositions based on PS, PMMA and styrene-acrylonitrile copolymer grafted to various fillers and found that it was possible, in this manner, to obtain particles with a grafted polymeric coating of sufficiently homogeneous thickness.

The important thing is that in all works where the uniformity of filler distribution in the matrix has been studied (cf., e.g. [316, 317]) a better dispersion of the filler has been reported for PFCM than could be achieved in the case of mechanical mixing of ingredients even if carried out in a solvent.

Using calorimetry to estimate the degree of filler-polymer interaction as described in [99] the authors of [318, 319] determined that the filler reaction with PVC is exothermic, which is indicative of a stronger bond in the polymer-filler system. No thermal effect was noted for mechanical mixtures, except for a few cases where it was endothermal.

2.3 Properties of Polymerization-Filled Composites

From the data obtained by different experiments ir may be concluded that in polymerization-filled composite materials (PFCM) a macromolecular layer is formed around each mineral particle strongly bonded to each either physically or chemically. The thickness of such a coat is much greater than that of the consolidated surface ("thin") layers in conventional filled composites [307]. The macromolecules constituting the further layers of polymer around the mineral particle in PFCM may not be in direct contact with the particle but may come into mechanical engagement with the firmly secured surface layer (for example, the macromolecules of different layers may belong to one and the same submolecular formation). The range of the perturbing effect of a filler particle should, therefore, be much more extended in PFCM than in mechanical mixtures of similar composition.

Spherical indentor deformability tests of composites with PE synthesized on different fillers have shown [164] that the deformation of PFCM containing high percentages of filler (and polymer concentrations of less than 10% by mass) is substantially elastic and the specimen recovers completely after release of the load. As the polymer content increased to 60% by mass considerable residual deformation was noted: the material becomes flowable and can be squeezed through the viscometer capillaries. With up to 80% by mass of polymer, the behavior of PFCM in viscometric experiments remained virtually unchanged, except for the characteristic flow separation which occurred at flow rates much smaller than normally employed in real-scale processing.

Technologically it is important to obtain PFCM by polymerization of moldable and extrudable ethylene grades. An attempt to solve this problem was reported in [320, 321]. The authors investigated ethylene polymerization in heptane in the presence of disperse fillers – kaolin, talc, aluminum hydroxide – activated with the $VOCl_3$-$AlEt_2Cl$ catalytic system. The molecular mass of the polymer was controlled with the help of hydrogen. In this manner is was possible to bring the MM of PE down from $(1.8-2.5) \times 10^6$ to $(1.2-1.6) \times 10^5$ and thus obtain materials characterized by melt flowability indices close to the extrusion grades. Figures 6, 7 show certain physico-mechanical characteristics of such composites as a function of the MM of PE. It is seen that reduction of MM had a determinetal effect on some physical characteristics. Table 10 contains selected strain/stress parameters of polyethylene composites synthesized on various fillers [320, 321]. The molecular mass of the polymer was $(1.2-1.6) \times 10^5$, filler content 28–30% by mass.

The unfavorable effect of reduced MM of polyethylene on the complex of properties has also been reported in [322, 323]; the parameter ε is the most affected. From Table 10 one can see that this characteristic can be improved considerably

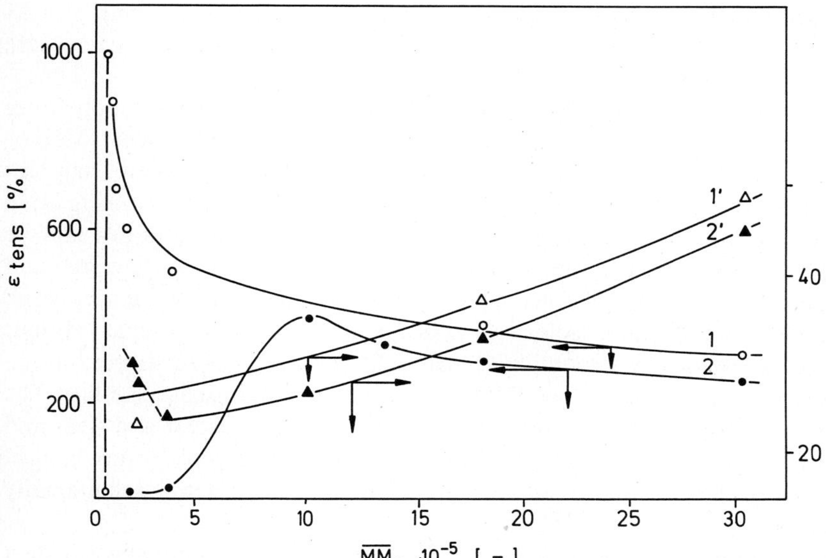

Fig. 7. Variation of relative elongation at rupture (ε_{tens}) and ultimate tensile strength (σ_{tens}) of the polymer matrix (I, I′) and polyethylene-based composites polymerization filled with kaolin (2,2′) in function of polymer MM [320]. Kaolin content 30% by mass. Conditions same as in Fig. 6

Table 10. The strain/stress characteristics of polymerization-filled composites [320, 321]

Characteristic	PE grade	PE filled with:					
		Kaolin[a]		Talc[b] Heat treatment		Aluminum hydroxide[c]	
		I	II	Anneal	Temper	I	II
σ_y, MPa	30.0	32.3	25.2	25.4	19.8	23.8	21.1
ε_y, %	8.5	2.7	2.9	2.8	4.4	5.3	5.3
σ_{tens}, MPa	23.4–30.5	32.3	24.7	24.7	16.0	19.6	16.4
ε_{tens}, %	600–750	2.7	3.9	3.4	20.50	100–300	100–300
E, MPa	1300	2600	2750	2600	1500	2000	1400

[a] I – the filler was activated in the liquid phase via reaction of the organoaluminum compounds fixed on the surface with $VOCl_3$; II – activation of filler by previous securing on the surface of $VOCl_3$ from the gas phase; particles of less than 5–10 microns in size amount to 96%;
[b] Particles of less than 20 microns make 90%;
[c] I – Particle size 2 to 50 microns; d_{aver} 8 microns; II – particles smaller than 4 microns – 20%; 4–10 microns – 25%; 10–20 microns – 36%; larger than 20 microns 15%.

by using an appropriate filler. This however tends to slightly reduce the rigidity and strength of the composites.

In [324] it was noted that in case of polyethylene-based PFCM it is the MM which is the parameter controlling the composite viscosity. For a fixed MM, the viscosity of the melt increases symbatically with the filling percentage by mass. The nature of the viscosity versus MM relationship is similar for both PFCM and mechanical mixtures, that is, in terms of rheology the two systems are practically identical.

Unless the MM is controlled, the non-flowing material results only from polymerization filling of PE. For other polyolefins, e.g. polybutene [325] or polypropylene [326, 327] the polymerization-filled materials will have sufficient flowability to enable their processing either by extrusion or pressure molding using conventional apparatus. The PFCM based on PE in the pure form can now be commercially processed only by such methods as caking, pressing, vibro- and strang-molding. In [328] it has been noted that polymerization-filled polyethylene powder is readily pressure molded without application of heat (briquetted) into dense and mechanically strong parts. The molded blanks can be sintered in the free state at a temperature above 150 °C with no detriment to shape stability. The molded polyethylene-based PFCM have a very low friction factor and wear and are superior as far as these characteristics are concerned not only to mechanical mixtures of identical composition but also to most of the materials traditionally used at friction joints [306, 328].

In [17] it is reported that PFCM, with PE synthesized on kaolin and having a MM of $>10^6$ and containing up to 50% by mass of filler, feature very high strain stress characteristics. For the composites containing 30 and 50% by mass of kaolin the value of σ_{tens} is 46.44 and 23 MPa respectively; ε_{tens} decreases from 260 to 240 and 150%, and the modulus increases from 900 to 1900 and 2900 MPa.

Table 11. Properties of heat insulation materials obtained by synthesis of PE on perlite [18]

Material	Polymer %	Compr. strength σ_{comp}, MPa	Bending strength σ_{bend}, MPa	Density kg/m^3	Heat conductivity coefficient, kcal/m s deg
PFCM: perlite-polyethylene	10–12	5–9	3	120–150	0.03–0.04
Perlite -phenol formaldehyde resin [a]	70	3–5	3	120–150	0.03–0.04

[a] The comparison is made with a heat insulating material which is commercially available.

The superhigh-filled composite materials obtained by PE polymerization on formed perlite particles present in the amount of 80 to 95% by mass can be used to advantage as heat insulation materials (see Table 11).

In [317] the properties of materials based on PS, PMMA, styrene-acrylonitrile copolymer (SAC) grafted to SiO_2 used as a filler of both the disperse and fibrous nature were studied. The authors discovered that complete fixation of PS on particle surfaces greatly increased the polymer rigidity: the value of E′ of the composition filled to 29% by mass increased in the range between −200 and +200 °C more than twice compared to simply mixed compositions. It is hypothesized that owing to formation of the pseudo-crosslinked structures as a result of overlapping of grafted polymeric layers belonging to individual filler particles, the macromolecular segments cannot make full advantage of the segmental mobility of the cooperative type. On further decrease of the grafted layer thickness (filling ratios of 65% by mass the composite behaved very much like a densely crosslinked polymeric system, obviously, for the reason of formation of a secondary network ("skeleton") of the filler.

Thermomechanical experiments [317] have shown that high-filled synthesized PS can stand considerable deformations at elevated temperatures. Just as in case of PE, the increase of polymer percentage in the composition improves resistance to large deformations. The maximum effect on the thermomechanical behavior of composites was noted for polymerization-filled rigid polymers with a strong interchain interaction. For PMMA it was noted, for example, that even a partial grafting of the matrix to the filler markedly decreased the strain in comparison with the compounded specimens. The PFCM with PP synthesized on different fillers feature a strength about equal to that of compounded samples, only the impact strength being exceptionally high against normal filled PP level: with 30% by mass of tufa the impact strength of PFCM was higher than that of unfilled material [326, 327]. Besides, the PFCM are reported to have a superior frost resistance (up to −45°).

On the basis of the recent investigations we can outline two possible ways of creating materials on the basis of polymerization-filled composites:

1) synthesis of material with the desired filler content and polymeric coat MM and its direct processing to articles;

2) the use as filler of polymerization-filled polymer particles with a high organic component content, the base polymer not necessarily being of the same nature as that of the coats in PFCM.

Materials of the second type can be classified as composites with polymerization-modified fillers (PMF) or filler concentrates.

The wear of the working parts of processing equipment is known to be one of the factors hampering the processing of filled polymers. The wear can be mechanical (abrasive, hydroabrasive, erosion), corrosion-mechanical, wear due to "seizing" [329]. The wear of the working parts and the resultant increase of the radial clearances between the ridges of screw conveyors and the extruder cylinder walls have a very detrimental effect on the machine operation lowering productivity and increasing the production cost. The wear can no longer be tolerated when the losses become comparable with the cost of replacing the worn screw with a new one. The replacement becomes economically feasible even with a 10% reduction of productivity [330].

We have already noted in an earlier section that treatment of filler with a finishing agent also increases its abrasive attack on the parts of the processing equipment [271]. At the same time it has been shown indirectly in [297, 298] and directly by microscopic inspection in [294] that a synthesized polymer film remains on the polymerization-modified filler after the processing treatment, that is, there is always a medium which protects the molding equipment from the abrasive attack of the filler. In view of these observations, the polymeric composites with PMF are comparable, in terms of abrasive activity, with unfilled polymers [226, 227].

2.4 Physico-Mechanical Properties of Composites with Polymerization-Modified Fillers (PMF)

In a series of investigations [293, 297, 298, 331] it was shown that the strength versus inorganic phase content relationship of composites with polyethylene-"clothed" fillers has a typical extremum. In practically the entire range of filler concentrations investigated (30 to 70% by mass) the strength of PMF composites was higher than that of the base polymer; the reinforcement was characterized by a factor of as high as 2 to 2.5 depending on the nature of the filler, otherwise present in an optimum concentration, its disperseness, polymer film thickness (normally the optimum filler concentration is within 30 to 50% by mass).

In [332] it was noted that the strength of samples cut out at different locations of an article made from filled thermoplastics by pressure molding may differ widely – which is due to the nonuniform orientation of the polymer at different locations of the mold. The very high strength parameters of composites with PMF in molded specimens are obviously also due to orientation effects, while for standard mixed samples of similar composition (that is, a matrix which, apart from the filler, contains some superhigh molecular polyethylene imitating the PMF coats) the

Table 12. Properties of filled PVC (filler: 60 parts by wt. of chalk to 100 parts by wt. of PVC) [316]

Filler	σ_{tens}	σ_{comp}	Durability under stress, S		ε %	E MPa	a MPa cm
	MPa		30 MPa	40 MPa			
Modified chalk	53	75	6.3×10^5	5.4×10^3	7	1210	0.18
Unmodified chalk	36	60	60	1.0	22	1090	0.35
Unfilled PVC	55	65	3.21×10^5	3.6×10^3	125	1050	0.31

strength is markedly lower even where the samples have been obtained under substantially identical conditions. The difference between composites with PMF and standard samples is also significant as regards the impact strength [164, 331]. Whatever the nature of the mineral filler, this index is 1.5–2.0 times greater for the first than for the second.

On the contrary, the impact strength is decreased in PVC-based samples filled with chalk modified by grafted PVC, as it is seen in Table 12 below.

One can see that the relative elongation is lower in composites, but the modulus and, what is particularly important, durability of samples is higher. Obviously, the interaction between the polymeric coat and the matrix is particularly important in materials of this kind. This is illustrated by the data of Table 13 below.

The authors of [210] used the measured enthalpy values (ΔH) to compare the compatibility of fillers coated with different materials with the base PVC by the procedure described elsewhere [99]. The results are given in Table 14.

From the table it is seen that the enhanced energy interactions between the polymeric coat of the filler and the matrix does not always entail an upgraded complex of physico-mechanical characteristics of the composite. The authors of [210] have advanced an opinion that the enhanced filler-matrix interaction prevents formation of labile bonds between the two and hinders the relaxation processes at the interphases.

It must be pointed out that polymerization-filled composites with modified fillers may behave in the manner described in [329] where 0.3–3% of PMMA was

Table 13. Effect of the chemical nature of the polymeric coat on properties of filled PVC (filler: parts by wt. of chalk to 100 parts by wt. of PVC) [316]

Coat material	σ_{tens}, MPa	σ_{comp}, MPa
Unmodified chalk (without polymer coat)	30	48
Polyacrylic acid	34	66
Acrylic acid/butacrylate copolymer	28	65
Acrylic acid/acrylonitrile copolymer	33	65
Acrylic acid/vinyl chloride copolymer	49	75
Unfilled PVC	52	65

Table 14. Thermal effect (ΔH) of interaction of chalk coated with different polymeric materials with the PVC matrix and the mechanical properties of compositions with 40% by mass of chalk [210]

Polymer coat	ΔT J/kg PVC	σ_{tens} MPa	σ_{comp} MPa	ε %	a kJ/m^2
–	+ 5.8	40	36	15	3.0
PVC	+ 3.7	55	53	9	2.0
Poly(butyl acrylate)	–14.3	40	33	30	3.2
Copolymer of vinyl chloride and 2-ethyl-hexylacrylate	–19.5	47	40	15	2.1

grafted to glass beads. Using scanning electron microscopy they showed that the grafted polymer chain became intertwined but did not mix with the homopolymer used as the matrix. In failure, the integrity of such a material is violated along the grafted polymer-homopolymer boundary. Behavior of this kind is particularly likely in systems with fillers coated with PE films because of the latter's very high molecular mass.

An attempt to correlate the coat grafting degree, its MM and the number of grafted chains with the degree of its interaction with the homopolymer and the physico-mechanical characteristics of compositions was undertaken in [333, 334]. Aerosil was treated with methylvinyl dichlorosilane which, as shown in [335], promotes formation of a strong chemical bond between the grafted polymer and Aerosil; the number of active sites generated on the surface of modified silica filler under radiation is directly proportional to the amount of exposure. Postradiation graft polymerization of butyl methacrylate was carried out. The data obtained in [333, 334] are listed in Table 15 below.

The enthalpy of interaction of the grafted polymer with the filler surface is given by

$$\Delta H = \Delta H_3 - (W_1 \Delta H_1 + W_2 \Delta H_2),$$

where ΔH_1, ΔH_2 and ΔH_3 are heats of mixing with chloroform of: 1 g of PMBA, 1 g of Aerosil with methylvinyl dichlorosilane finish, 1 g of polymerization filled PMBA, respectively.

Note that the variation of the enthalpy of filler interaction with the coat (ΔH) is fully determined by ΔH_3, since ΔH_1 and ΔH_2 remain practically constant.

The conclusions from the table data are easy to see. We would only note that, as reported in [333], in specimens with a different number of grafted chains one first observes a symbatic increase of tear strength and elasticity modulus with the increasing degree of grafting and MM. Yet, on further increase of the degree of grafting these parameters pass through a maximum and then go down. This behavior can, according to the authors of [333] be due to the increase of adhesion with the degree of grafting on the one hand; on the other, the higher grafting degree should cause loosening of the grafted chain packing.

Table 15. Physico-mechanical characteristics and heats of mixing with chloroform (ΔH_3) of poly(butyl methacrylate) polymerization-filled with aerosil (10% by mass) [333, 334]

$N^* \times 10^4$ mol/kg of filler	q*, %	$\bar{M}_\eta \times 10^{-6}$	$\Delta H_3 \times 10^{-3}$	E MPa	σ_{tens} MPa	ε %
PMBA	–	–	–	61	9.5	150
PMBA + unmodified Aerosil	–	–	–	138	12.7	156.5
PMBA + modified Aerosil	–	–	–	164	14.2	121
0.23	4.0	1.75	+ 6.4	203	14.2	103
	5.9	2.57	+.36	209	14.2	130
	6.9	2.96	–	248	15.0	112
	12.6	5.43	– 7.3	377	15.7	130
0.72	7.4	1.01	– 0.8	249	14.4	117
	8.9	1.52	–	272	14.7	116
	19.6	2.57	–	371	17.0	111
	26.9	3.88	–13.8	339	15.9	123
	30.2	4.07	– 6.5	235	15.5	143
1.55	13.5	–	–	247	15.6	83
	20.7	1.42	– 3.6	245	10.5	104
	30.7	1.86	– 5.8	201	13.4	108
	36.6	2.48	– 8.3	180	12.4	134
	55.3	3.46	–11.4	–	–	–

Notes: * q is the degree of grafting; N is the number of grafted PMBA chains.

For composites with polymerization-modified filler it is typical that the physico-mechanical characteristics should increase symbatically with the quantity of polymer which becomes attached to the filler in the polymerization process. This effect has been observed for polyethylene [293, 321], poly(vinyl chloride) coats [316], and in [336, 337] for kaolin coated with poly(vinyl acetate) and introduced into the copolymer of ethylene and vinyl acetate.

2.5 Rheological Properties of Composites with Polymerization-Modified Filler (PMF)

It has been found that, for a fixed mineral filler content, the viscosity of PMF-based composites increases when the coat is made of polyethylene [164, 209, 293], poly(vinyl chloride) [316] and polypropylene [326, 327]. The picture was different, however, for composites based on the ethylene/vinyl acetate copolymer to which kaolin with grafted poly(vinyl acetate) was added [336]. Addition of PMF with a minimum quantity of grafted polymer results in a sharp drop of flowability (rise of viscosity), in comparison to addition of unmodified filler; but with a further increase of the quantity of grafted polymer the flow gradually increases and, depending on the kaolin content and quantity of grafted polymer, may even become higher than in specimens with unmodified filler, for equal concentrations.

The flow of polymerization-filled composites is, as shown in [338] for composites with chalk coated with copolymer of methyl methcrylate and methacrylate, also largely dependent on the composition of the copolymer and sharply decreases when the percentage of methyl methacrylate in the copolymer increases.

In HDPE and LDPE matrices, the introduction of PMF with polyethylene coats affects the viscosity in such a way that is independent of the nature and particle size distribution of the mineral filler employed and depends only on NMF particle size. All other conditions identical, this effect is the same as that of the particles of standard high-molecular PE [164, 297, 298].

In a number of works (e.g. [339–341]) the authors sought to superimpose graphically the flow curves of filled melts and polymer solutions with different filler concentrations; however, it was only possible to do so at high shear stresses (rates). More often than not it was impossible to obtain a generalized viscosity characteristic at low shear rates, the obvious reason being the structurization of the system.

For PMF-filled polymer melts, a marked tendency to form the so-called secondary network has been reported in a number of works. For PE based composites it has been shown, for example, in dynamic loading tests that at loading frequencies tending to zero, the limiting complex modulus of elasticity (a counterpart of the yield limit) may, at fixed inorganic phase percentage in PMF compositions, be several orders of magnitude higher than that in systems with unmodified filler [342]. Characteristically, a linear correlation has been reported [342] between the limiting dynamic shear modulus of melt and compressive strength of material. Earlier it was also shown for ionol/(unmodified and modified with grafted PMMA) perlite systems that the yield limits of melts with PMF is far greater than that of systems with unmodified perlite [70].

It is possible, however, to obtain generalized concentration-invariant curves under straining rates close to those used in the real-scale commercial process. In [163, 164, 209] the generalized curves for PE based composites were obtained by the procedure described in [340]: by carrying out nonrotational shifts in the vertical and horizontal directions the authors sought to achieve the closest coincidence between the experimental curves in the $\lg \eta - \lg \tau$ coordinates for the base polymer and the curves for filled composites.

In [343] it was shown, with regard to the hydrodynamic effect of fillers on the viscoelastic properties of composites, that the dynamic functions must obey the following equations:

$$G'_c(\omega)\,[G_m(\omega)]^{-1} = G''_c(\omega)\,[G_m(\omega)]^{-1} = b(c)\,, \tag{39}$$

where ω is frequency.

This means that the dynamic functions of composites can be reduced to the dynamic functions of the matrix by a mere shift along the modulus axis by b(c).

The authors of [344] investigated PS based composites prepared by compounding unmodified Aerosil with Aerosil to which a PS layer was grafted. The MM of the matrix and the grafted layer was 3×10^5. The resultant composites are characterized in Table 16 below.

Table 16. Composition[a], frequency (ω) beginning from which one may achieve coincidence of the G′ values and the shift factors b(c) at 190 °C [344]

Composition No.	Free PS, %	Bound PS, %	Filler (Aerosil) %	Frequency ω s^{-1}	Shift factor b(c) used to reduce G′(ω)
1	100/100	0/0	0/0	–	–
2	80/89, 66	0/0	20/10, 34	5	9.12
3	69, 23/77, 59	10, 77/12, 07	20/10, 34	25	9.77
4	60/76, 50	0/0	40/23, 50	56	70.8
5	39, 46/49, 04	21, 54/27, 46	40/23, 50	95.5	95.5

[a] The numerator gives filling ratios by mass, denominator – by volume.

From the results obtained in [344] it follows that the composites with PMF are more likely to develop a secondary network and a considerable deformation is needed to break it. As the authors of [344] note, at low frequencies the G′(ω) relationship for Specimens Nos. 4 and 5 (Table 16) has the form typical of a viscoelastic body. This kind of behavior has been attributed to the formation of the spatial skeleton of filler owing to the overlap of the "thin" boundary layers of polymer. The authors also note that only plastic deformations occurred in shear flow.

At low shear stresses the viscosity of PS based compositions with PMF is higher than that of compositions with unmodified filler; the higher the shear stresses the closer the respective viscosities.

Under stress conditions similar to those arising in the extrusion or pressure molding processes, the activation energy for viscous flow of systems with PMF is little different from that for the matrix [163, 164, 209, 344, 345]. This means that the "secondary" network of the material has been destroyed [69].

3 Some Special Effects in the Flow of Filled Polymers

1. One sometimes observes what may be called "migration" of the filler, whatever its particle shape, in suspensions flowing along the extruder channels. The direction of the "migration" depends on the properties of the dispersed phase and the matrix [346]. Radial (relative to the flow axis) motion of filler particles was observed both in low-molecular liquids [347, 348] and filled polymeric systems [349]. In viscometric experiments the depletion of filler in the surface layers can manifest itself phenomenologically as an enhanced anomaly of viscosity. The authors of [350] also observed a nonuniform distribution of filler along the direction of injection of PP melt filled with glass balls into the spirally-shaped mold. The filler concentration was maximum at the end of the spiral and minimum at its beginning.

A thorough study of the distribution of filler in the molded specimen was carried out by Helger and Mennig [352]. The authors found that the filler distribution in the specimen depended on the particle size and shape and filler concentration, the materix properties and molding conditions.

Wu [353] who studied extruded fiber-filled composite samples established that the distribution of fibers along the radius of the specimen depended on the extrusion rate: at low rates the fiber distribution is uniform, at medium rates the minimum of filler concentration occurs at $\simeq 0.63$ R (R is the extrudate radius) (in case of spherical particles this point corresponds to the maximum filler concentration); at high extrusion rates most of the fibers are concentrated about the flow axis and there are almost none on the extruded sample surfaces.

For anisodiametric fillers, one observes a different orientation of the filler across an extruded or molded specimen, also dependent on the rate of injection. At very high extrusion rates the flow of filled material changes from the volumetric to what may be called the "plug" type, which must also affect the filler orientation. In a number of works, (for example [176, 349, 354–356]) where high injection rates were used the extrudate usually had peripheral layers with random filler orientation followed by layers with a high degree of orientation along the flow and, finally, in the core the filler is orientated virtually perpendicularly to the flow axis. It has been noted that the layers characterized by a high degree of filler orientation in the direction of flow become thicker as the temperature is increased.

The eccentric rotation of filler in virtue of the variant shear rates along the extrudate cross-section has been proposed as an alternative to the plug flow mechanism to explain this orientation pattern [355]. In [357] it was noted that the cross-wise orientation of fibers at the core of molded specimens was established only after the entire mold had been filled; the authors assumed this orientation to be due to the quasi-static stresses arising in the material under pressure.

2. Another well-known effect which accompanied the flow of systems with a fibrous filler consists in that a filler "plug" is formed in front of the die or measuring instrument, through which the matrix filters forward [173, 189, 211]. A similar effect has been reported for dispersed fillers as well [208, 209]. Typically no filler "package" is formed at the capillary inlet for systems with PMF [208, 209]. The lower tendency for filler segregation in PVC based composites with PMF allows the extruder apparatus to work with a greater capacity in comparison with processing of simple mixtures if the filler content is the same [210, 316]. Improved processability has been noted for other PMF systems too [358].

3. For all high-filled polymers, whatever the nature of the matrix and filler, it has been found [297, 298] that the Bagley pressure loss at extruder inlet increases sharply in a certain range of temperatures and rates of deformation. What distinguishes composites with PMF is that for them the pressure loss at the inlet begins at lower deformation rates.

4 Some Recommended Practices in Processing Highly-Filled Compositions

Considering that the filler increases matrix viscosity, it is necessary to use processing equipment of high power permitting the development of higher extrusion and molding pressures in comparison with conventional apparatus. Whereas generally the filler suppresses the highly elastic behavior of melts, for example it reduces the probability of the so-called elastic turbulence [1], in filled polyolefins, PVC and some other polymers it gives rise, or enhances, the tendency for flow separation, the effect being the more important the higher the MM of the polymer. This must be taken into account when designing the process for such composites, since as the critical conditions develop the material may begin to slip against the wall of the material cylinder of the single-screw extruder, whereby the extrusion may be slowed down or even stopped. This problem is dealt with by employing of means of proving a "stiff" supply of material into the molding apparatus, e.g. a twin-screw extruder, or a single-screw machine provided with a special slotted sleeve in the feed zone. In choosing the appropriate processing equipment one should remember that it may be subject to the abrasive attack of the filled materials. For example, the type and degree of wear of the screws of a twin-screw extruder depends largely on their relative directions of rotation. The wear is smaller when the rotation is unidirectional [361].

The wear of screws can be reduced by increasing the clearances between the components of the working organs; it is most advantageous to increase the clearance between the screws than between each screw and the cylinder wall [362]. Good results are obtained by e.g. subjecting the working surfaces of the equipment to hardening treatment [363] or covering them with special grades of hard alloys [364]. Since the wear increases with increasing pressure [361], the latter should be as low as possible. This is also important from the power saving standpoint. One of the ways to solve this problem is to facilitate material movement through the instrument by using Teflon [211]. Yet, the development of a complex stressed state in the melt during extrusion by causing the apparatus parts to vibrate with sonic or ultrasonic frequency appears to be a more promising method [365].

Earlier we have said that the flow of filled materials can be raised by modification. One of the modification alternatives consists in using a combination ("flexible") filler. As shown in [366], addition of a small quantity of glass balls to composites filled with chalk caused a sudden increase of the flow of the system, provided the filler concentration is below critical. Similar behavior was also observed in fiber-filled systems when a small quantity of dispersed particles were added [33].

At fixed filling ratio, using a hybrid filler may decrease extruded material swelling, raise the critical deformation parameters at which elastic turbulence of the melt may begin [366].

The orientation of an anisodiametric filler which results from pressure molding is known to bring about anisotropy of properties in different directions, the extent of which is determined by the matrix polymer [367]. This is not always a plausible result from the technological standpoint, if only because orientation, frequently nonuniform, may build up to cause buckling of articles [368–370].

---

[1] "Elastic turbulence" here means the distortion of the surface evenness and shape of the extruded article. A review of the wealth of experimental data and reasons for this phenomenon is outside the scope of this paper. Detailed information concerning this problem will be found in [359, 360].

So far there is no "cover-all" recipe for remedying these undesirable phenomena; yet recommendations have been made for some special cases which included proper choice of process conditions and methods, rational design of molds and articles allowing one to minimize the negative effect of filler and matrix orientation in the manufacturing process [370].

Another important thing which may happen in the course of the process is the breakdown of the filler. A detailed study of situations of this kind for glass fiber filled composites has recently been conducted by Von-Turkevich and Erwin [371]. The authors concluded in favor of independence of the final length of fiber from filler concentration in the system; the tendency for fiber breakage decreases with the increasing process temperature (or decreasing material viscosity). Reprocessing has a particularly detrimental effect on the fiber status. Remarkable, the authors of [372] obtained contrary data according to which the comminution of the fibers is promoted with the increasing filler content in the processed material. Optimization of the technological process as shown, for example, in [373] may help minimize fiber breakage and, as consequence, improve the strength of the composite.

Table 17 contains data on the strength of mica-filled PP compositions.

It is seen that prior to processing, the strength characteristics are higher in materials with large particles, but they suffer greater damage during the process.

In order to keep all the advantages of the mica-filled composites it is necessary to so adjust the manufacturing process as to minimize its crushing. Pressing may be the appropriate technique, as it follows from the data of Table 18.

Yet pressing is an all too unproductive method and cannot be used as the main manufacturing process. Therefore special measures are to be taken to minimize the crushing of filler in composite filled with lamellar or fibrous materials.

Last but not least, destruction of the matrix may result from the shaping process, the extent of which depends primarily on the manufacturing process, conditions and type of filler [376, 377]. The destruction of the matrix in a composite may be

Table 17. Effect of the number of processing cycles on the properties of molded PP samples filled with mica particles [374]

Mean particle size, microns	Parameter	Value of parameter after the following number of processing cycles		
		0	3	6
30	Tensile modulus,	7.6	7.2	6.6
225	GPa	7.9	7.8	4.8
30	Compressive strength,	32	32	26
225	MPa	25	29	27
30	Bending modulus,	7.2	6.7	6.6
225	GPa	7.5	5.5	4.8
30	Bending strength,	63	63	79
225	MPa	54	59	68

Table 18. Effect of molding method on mechanical properties of mica-filled thermoplastics ($V_f = 0.5$) [375]

Thermoplastic	Pressing		Pressure molding		Extrusion followed by pressure molding	
	E_{bend} GPa	σ_{bend} MPa	E_{bend} GPa	σ_{bend} MPa	E_{bend} GPa	σ_{bend} MPa
ABC-plastic	41.6	154	36.9	105	32.5	99.9
Polystyrene	41.3	123	36.9	114	25.4	67.6
Polypropylene	37.9	172	26.9	86.1	–	–

much more severe than that which may result from processing the pure polymer. At the same time a lower degradation of the polymer may be expected, according to preliminary data [378–380], in composites with PMF.

5 References

1. Lipatov YuS, Denisenko PV (1985, 1986) Polymeric Composite Materials 26: 3; 29: 17
2. Reutov SV, Golev YuV (1976) Plast. massy 11: 22
3. Yakovlev MF, Ozerov GM (1981) Plast. massy 1: 54
4. Aminabhavi TM, Patel PC, Biradar NS (1981) Polym. Eng. Sci. 21: 1085
5. Aminabhavi TM, Biradar NS, Holeunavar RH (1983) J. Macromol. Sci. A20: 515
6. Piperov NL, Andrianov CA et al. (1987) Fiz. Khim. Mekhanika 15: 58
7. Solomko VP (1980) Filled Crystallizable Polymers Naukova Dumka, Kiev
8. Mascone CF (1985) Chem. Eng. 27: 27
9. Brown DS (1986) Plast. World 1: 67
10. Richardson MOV (ed) (1977) Polymer engineering composites Applied Science Publishers, London, Ch 9–13
11. Seymor R (1986) Pop. Plast 11: 43
12. Manson JA, Sperling LZ (1976) Polymer blends and composites Plenum Press, London, Ch 13
13. Enikolopyan NS (1978) Sz. Vses. Khim. 23: 243
14. Lipatov YuS (1984) Polymer Composites: A Prospective View Naukova Dumka, Kiev
15. Strength and Failure of Structures made from Composite Materials. Bibliography of home and foreign literature covering the period of 1951–1981 (1982) Novosibirsk
16. Enikolopov NS, Volfson SA (1978) Plast. massy 1: 39
17. Diachkovsky FS, Novokshonova LA (1984) Usp. Khimii 53: 200
18. Enikolopyan NS, Diachkovsky FS, Novokshonova LA (1982) In: Complex organometallic catalysts for olefin polymerization Publication IKRF 9: 174
19. Diachkovsky FS (1987) In: Reactions in polymer systems, Khimiya, Leningrad, p 174
20. Ivancheev SS, Dmitrienko AV (1982) Usp. Khimii 51: 1178
21. Dmitrienko AV, Ivancheev SS (1987) In: Reactions in polymer systems, Khimiya, Leningrad, p 201
22. Bryk TM, Lipatov TE (1986) Physics and chemistry of multicomponent polymer systems, Haukova Dumka, Kiev, p 9
23. Stamhius JE, Loppe JPA (1982) Rheol. Acta 21: 103
24. Nielsen LE (1974) Mechanical properties of polymers and composites, Marcel Dekker, New York, Ch 7

25. Nielsen LE (1957) J. Compos. Mater. 1: 100
26. Mills NJ (1971) J. Appl. Polym. Sci. 15: 2791
27. Nielsen LE (1968) J. Mater. Sci. 2: 120
28. Tolstaya SN (1982) Polymer Composite Materials 15: 6
29. Lewis TB, Nielsen LE (1968) Trans. Soc. Rheol. 12: 21
30. Tovmasyan YuM, Topolkarev VA, Oshmyan VG et al. (1985) Dokl. AN SSR 283: 681
31. Boaira MS, Chaffey CE (1977) Polym. Eng. a. Sci. 17: 715
32. Bretas RES, Powel RL (1985) Rheol. Acta 24: 69
33. Dreval VE, Tsidvintseva MN et al. (1987) Mekh. Komp. Mater. 3: 505
34. Ramsteiner F, Theysonn R (1984) Composites 15: 161
35. Pakharenko VA, Petrushenko EF et al. (1984) Plast. massy 1: 61
36. Solomko VP (1974) Mekh. Polimerov 2: 197
37. Mooney M (1951) J. Colloid. Sci. 6: 162
38. Landel RF, Moser BG, Bauman AJ (1965) Proc. 4th Congr. on Rheology, 4P2: 663 Interscience Publ. N.Y.
39. Maron SH, Pierce PE (1956) J. Colloid. Sci. 11: 80
40. McCeary RL (1961) J. Amer. Ceram. Soc. 44: 513
41. Farris RT (1968) Trans. Soc. Rheol. 12: 281
42. Chong TS, Christiansen EB, Bayer AD (1971) J. Appl. Polym. Sci. 15: 2007
43. Ferrigno TH (1978) Polym. Eng. Sci. 18: 33
44. Milewsky V (1978) In: Handbook of fillers and reinforcements for plastics, Van Nostrand Rheinhold, New York, Ch. 3
45. Ismailov TM, Sagolaev GV (1968) Plast. Massy 4: 44
46. Onogi S, Matsumoto T, Warashina J (1973) Trans. Soc. Rheol. 17: 175
47. Prokopenko VV, Titova OK et al. (1977) Vysokomol. Soed. A19: 95
48. Kargin VA, Sogolova TI, Metelskaya TK (1962) Vysokomol. Soed. 4: 601
49. Cuzeev VV, Rafikov MN, Malinsky YuM (1975) Vysokomol. Soed. A17: 804
50. Lutsky MS, Klitenik GS, Fridman ID (1977) Kolloid. Zhurnal 39: 376
51. Malinsky YuM (1970) Usp. Khimii 39: 1511
52. Lipatov YuS (1980) Interphase effects in polymers, Naukova Dumka, Kiev
53. Lipatov YuS (1977) Physical chemistry of filled polymers, Khimiya, Moscwa
54. Lipatov YuS (1984) Colloid chemistry of polymers, Naukova Dumka, Kiev
55. Theocaris PS, Sideradis E (1984) J. Appl. Polym. Sci. 29: 2947
56. Theocaris PS (1985) Adv. Polym. Sci. 66: 150
57. Eiglis MP, Moiseev VM (1972) In: Surface forces in thin films and disperse systems, Nauka, Moscow, p 271
58. Rusanov AN (1973) In: Advances in colloid chemistry, Nauka, Moscow, p 39
59. Deryagin BV (1973) In: Advances in colloid chemistry, Nauka, Moscow, p 30
60. Privalko VP, Lipatov YuS et al. (1971) Vysokomol. Soed. A13: 103
61. Theocaris PS (1985) Colloid. Polym. Sci. 263: 863
62. Malinsky YuM, Epelbaum IV et al. (1966) Vysokomol. Soed. 8: 1886
63. Bakhtrakh GS, Malinsky YuM (1969) Kolloid. Zhurnal 31: 9
64. Bartenev GM, Zakharchenko NV (1962) Kolloid. Zhurnal 24: 121
65. Rowland F, Bulas R (1965) Ind. Eng. Chem. 9: 46
66. Bakhrakh GS, Malinsky YuM (1973) Kolloid. Zhurnal 35: 431
67. Lipatov YuS, Privalko VP, Shumsky VF (1973) Vysokomol. Soed. A15: 2106
68. Lipatov YuS, Todosiychuk TT et al. (1973) Vysokomol. Soed. A15: 2243
69. Faitelson LA, Alekseenko AI (1976) Mekh. Polimerov 3: 478
70. Iwakura K, Fajiumura T (1979) J. Appl. Polym. Sci. 24: 975
71. Nagashiro W, Tsunoda T (1980) J. Appl. Polym. Sci. 25: 2961
72. Kuznetsov GK, Irgen LA (1976) In: Thermodynamic and structural characteristics of polymer interphases, Naukova Dumka, Kiev, p 94
73. Kuznetsov GR, Irgen LA (1973) Mekh. Polimerov 3: 481
74. Kuznetsov GR, Irgen LA (1974) Mekh. Polimerov 2: 252
75. Kuznetsov GR, Irgen LA, Chirkova EA (1974) Mekh. Polimerov 3: 113
76. Rafikov MN, Guzeev VV, Malysheva GP (1971) Vysokomol. Soed. A13: 2625

77. Pervak IG, Kudypina TA (1977) Plast. Massy 8: 19
78. Goldman AYa, Khaparin VS, Sysoev IV (1986) Mekh. Kompoz. Mater. 3: 392
79. Malinsky YuM (1979) In: Problems of polymer composites, Naukova Dumka, Kiev, p 16
80. Kwei TK, Kamins CA (1964) J. Appl. Polym. Sci. 8: 1483
81. Lipatov Yu, Moysya E, Semenovich G (1975) Polymer 16: 582
82. Lipatov YuS, Privalko VP (1977) Vysokomol. Soed. A19: 125
83. Shanks RA (1986) Brit. Polym. J. 2: 75
84. Lipatov YuS, Sergeeva LN (1972) Adsorption of polymers, Naukova Dumka, Kiev
85. Lipatov YuS, Privalko PV (1985) Dokl. AN SSSR 284: 651
86. Semenovich GM, Lipatov YuS (1986) In: Physical chemistry of multicomponent polymer systems. Naukova Dumka, Kiev, p 186
87. Lipatov YuS, Shipov VV (1981) Kompoz. Polim. Mater. 11: 55
88. Bogdanova LM, Irzhak VI, Rozenberg BA (1984) Mekh. Kompoz. Mater. 3: 554
89. Rozenberg BA, Irzhak VI (1986) In: Polymer Composites and their Use in National Economy. Fan Press, Tashkent, p 178
90. Simonov-Emelyanov ID, Yakhnin ED et al. (1971) Plast. Massy 10: 41
91. Veĉerka F, Horak Z (1980) Proc. of 8th Int. Congr. Rheol. 7P2: 549 Plenum press, New York
92. Hirai N, Eyring H (1959) J. Polym. Sci. 37: 1
93. Gudova EG, Lipatov YuS, Todosiychuk TT (1987) Vysokomol. Soed. B29: 539
94. Lipatov YuS, Fainerman AE, Anokhin OV (1976) Dokl. AN SSSR 231: 381
95. Chirkova EA, Irgen LA, Eltekov YuA (1974) Mekh. Polimerov 3: 526
96. Lipatov YuS, Privalko VP et al. (1987) Vysokomol. Soed. B29; 163
97. Lipatov YuS, Khramova TS et al. (1988) Vysokomol. Soed. A30: 443
98. Lipatov YuS, Privalko VP (1972) Vysokomol. Soed. A14: 1643
99. Bessonov YuS, Tager AA et al. (1979) Vysokomol. Soed. A20; 99
100. Yushkova SM, Guzeev VV et al. (1985) Vysokomol. Soed. 25: 2109
101. Guzeev VV, Martynova LM et al. (1980) Plast. Massy 10: 32
102. Lipatov YuS, Privalko VP (1984) Vysokomol. Soed. B26: 257
103. Burton R, Folkes MJ (1983) Plast. a. Rubb. Proc. a. Appl. 2: 129
104. Folkes MJ, Hardwick ST (1984) J. Mater. Sci. Lett. 3: 1071
105. Vnukova VG, Kiselev VYa et al. (1984) Vysokomol. Soed. B26; 733
106. Malinsky YuM, Titova NM, Aleshidze MP (1973) Vysokomol. Soed. B14: 485
107. Malinsky YuM, Titova NM (1976) Vysokomol. Soed. B18: 259
108. Maurer FHJ, Kosfeld R (1985) Th. Uhleubroich. Colloid a. Polym. Sci. 263: 624
109. Privalko VP, Lipatov YuS, Khmelenko GT (1975) Polymers: Synthesis and physical chemistry 16: 107
110. Mamunya EP, Lebedev EV, Demchenko SS (1986) Kompoz. Polim. Mater. 30: 58
111. Kulernev VN (1983) Polymer Blends. Khimiya, Moscow
112. Malinsky YuM, Titova NM, Sakeev NF (1975) Vysokomol. Soad. A17: 1600
113. Malinsky YuM (1979) In: Problems of polymer composite materials, Naukova Dumka, Kiev, p 16
114. Carton A, Kim SW, Willes DM (1982) J. Polym. Sci. Polym. Lett. Ed. 20: 273
115. Khozin VG, Murafa AV, Cherevazhsky AM (1987) Mekh. Komp. Mater. 1: 130
116. Babich VF, Lipatov YuS (1968) In: Physico-chemistry of polymer multicomponent systems, Naukova Dumka, Kiev, p 222
117. Lipatov YuS, Babich FV, Korzhuk NJ (1985) In: Molecular characterization of composite interphases, Plenum Press New York, p 237
118. Sagalaew GW, Tourne M, Bischof C (1977) Plast. und Kautschuk 24: 681
119. Papanicolaou GG (1978) Colloid a. Polym. Sci. 256: 625
120. Papanicolaou GG, Theocaris PS (1979) Colloid and Polym. Sci. 257: 239
121. Papanicolaou GG, Theocaris PS, Spathis CD (1980) Colloid a. Polym. Sci. 258: 1237
122. Theocaris PS, Spathis G, Kefales B (1982) Colloid, a. Polym. Sci. 260: 837
123. Theocaris PS, Varias AG (1987) Rheol. Acta 1987, 184: 322
124. Lipatov YuS, Babich VF, Perepelitsyna LN (1982) Vysokomol. Soed. 24D: 548

125. Lipatov YuS, Babich VF, Svyatenko GP (1983) Kompoz. Polim. Mater. 19: 65
126. Babich VF, Perepelitsyna LN, Lipatov YuS (1984) Kompoz. Polim. Mater. 20: 15
127. Lipatov YuS, Babich BF (1987) Mekh. Komp. Mater. 1: 17
128. Takayanagi M, Uemura S, Minami S (1964) J. Polym. Sci. PC5: 113
129. Kanaun SK, Goldman AYa, Kudryavtseva LT (1986) Mekh. Kompoz. Mater. 6: 1093
130. Gandelsman MI, Matveenko RV, Budtov VP (1986) Kompoz. Polim. Mater. 30: 42
131. Solomatov VI, Bobryshev AN, Proshin AP (1982) Mekh. Kompoz. Mater. 6: 1008
132. Alter H (1965) J. Appl. Polym. Sci. 9: 1525
133. Richardson GC, Sauer JA (1976) Polym. Engng. a. Sci. 16: 252
134. Bramusso M, Savadori A, Bocci D (1985) Polym. Compos. 1: 1
135. Bhattchariya SK (1984) Polym. Commun. 25: 10
136. Svetlova V (1987) Angew. macromol. Chem. 153: 197
137. Bustedt J, Erlandsson AM et al. (1987) J. Vinyl Technol. 3: 136
138. Sumita M, Shirama T et al. (1983) J. Mater. Sci. 18: 1758
139. Sumita M, Tsuakumo Y et al. (1983) J. Macromol. Sci. Pol. Phys. Ed. B22: 601
140. Sumita M, Shirama T et al. (1984) Rheol. Acta 24: 396
141. Sumita M, Ookuma M (1984) Colloid a. Polym. Sci. 262: 103
142. Somov AI, Chernykh OV (1967) Thermal Stability of Materials and Structural Components 4: 76
143. Chow, TS (1982) J. Polym. Sci. Polym. 20: 2103
144. Golovkin GS (1974) In: Structural plastics. Khimiya, Moscow, p 11
145. Kelly A, Tyson WR (1965) J. Mech. Phys. Solids 13: 329
146. Kelly A, Tyson WH (1965) In: VF Zachay (ed) High strength materials, Wiley, New York, Ch 3
147. Tarnopolsky YuM, Skudra AM (1966) Structural strength and deformability of glass reinforced plastics. Zinatne, Riga
148. Tamuzh VP, Azarova MT et al. (1982) Mekh. Kompoz. Mater. 1: 34
149. Prut EV, Kompaneets LV et al. (1986) Mekh. Kompoz. Mater. 6: 1103
150. Korabelnikov YuG, Tamuzh VP et al. (1984) Mekh. Kompoz. Mater. 2: 195
151. Korabelnikov YuG, Siluyanov OF et al. (1987) Mekh. Kompoz. Mater. 2: 270
152. Ovchinsky AS, Sakharova E (1981) Mekh. Kompoz. Mater. 4: 620
153. Bowyer WH, Bader MG (1972) J. Mater. Sci. 7: 1315
154. Verkhovets AP, Utevsky LR et al. (1980) Mekh. Kompoz. Mater. 4: 740
155. Mitta RK, Gupta VB (1982) J. Mater. Sci. 17: 1315
156. Hasskell WE, Petrie SP, Lewis W et al. (1983) Polym. Eng. Sci. 23: 771
157. Mittal J, Gupta VB, Sharma P (1987) J. Mater. Sci. 1987, 22: 1949
158. Xavier SV, Sharma YN (1986) Polymer. Compos. 1: 42
159. Woodhams RT, Xanthos M (1978) in: Handbook of Fillers and Reinforcement for plastics. Van Nostrand Rheinhold, New York. Ch. 19
160. Lackdawala K, Salovey R (1985) Polym. Engng. Sci. 25: 797
161. Lee MCH (1985) Polym. Engng. Sci. 25: 909
162. Suetsugu Y, White JL (1983) J. Appl. Polym. Sci. 28: 1481
163. Popov VL, Fridman ML (1981) Rheological and engineering properties of filled polymeric materials. NIITEKhim, Moscow
164. Popov VL (1982) Rheological properties of norplastics and composites on their basis. Candidacy thesis. Institute of Chemical Physics AN SSSR publication, Moscow
165. Iwakura K, Fujimura T (1977) Kobunshi Ronbunshi 34: 111
166. Yanczak U (1980) Angew. macromol. Chem. 89: 47
167. Yanchak VM (1981) Vysokomol. Soed. B23: 7
168. Tolstaya SN (1979) Problems of polymer composite materials, Naukova Dumka, Kiev, p 3
169. Vinogradov GV, Malkin AYa (1980) Rheology of polymers, Mir Press, Moscow, Ch 6
170. Utracki LA (1983) Rubb. Chem. Technol. 57: 507
171. Maschmeyer RO, Hill CT (1974) Adv. Chem. Ser. 134: 95
172. Murty KN, Modlen CF (1972) Polym. Engng. Sci. 17: 848
173. Machmeyer RO, Hill CT (1977) Trans. Soc. Rheol. 21: 195

174. Czarnecky L, White JL (1980) J. Appl. Polym. Sci. 25: 1217
175. Malkin AYa, Epple TV, Gritsuk AI (1972) Colloid. Zhurnal 34: 550
176. Growson RJ, Folkes MJ, Bright PF (1980) Polym. Engng. Sci. 20: 925
177. Suto S, Fujimura T (1982) Polym. Engng. Sci. 22: 197
178. White JL, Charnecky L, Tanaka H (1980) Rubb. Chem. Technol. 23: 823
179. Lyubcheva M, Iliev S, Kyubchev L (1980) Fiz. Khim. Mekhanika 15: 69
180. Kitano T, Kataoka T, Nagatsuka Y (1984) Rheol. acta 24: 20
181. Wang KJ, Lee LI (1987) J. Appl. Polym. Sci. 33: 431
182. Minogawa N, White JL (1976) J. Appl. Polym. Sci. 20: 501
183. Volfson S, Karp MG et al. (1985) Mekh. Kompoz. Mater. 6: 1086
184. Faitelson LA, Yakobson EE (1987) Mekh. Kompoz. Mater. 2: 358
185. Pokrovsky VI (1986) Kolloid. Shurnal 30: 881
186. Newman S, Trementorri QA (1965) J. Appl. Polym. Sci. 9: 3071
187. Growson RJ, Folkes MJ (1980) Polym. Engng. Sci 20: 934
188. Cotten CR (1968) Rubber Age 11: 51
189. Fridman ML, Yanovsky YuG (1975) In: Rheology of polymer and dispersed systems and rheophysics. Minsk, p 189
190. Agarwal PK, Bagley EB, Hill CT (1978) Polym. Engng. Sci. 18: 282
191. Nishimura T, Kataoka T (1984) Rheol. Acta 23: 401
192. Philippov BW, Gaskins FZ (1958) Trans. Soc. Rheol. 2: 263
193. White JL, Crwoder JW (1974) J. Appl. Polym. Sci. 18: 1013
194. Faulkner DL, Schmidt LR (1977) Polym. Engng. Sci. 17: 657
195. Péna JJ, Gusnan GM, Santamaria A (1977) Polym. Engng. Sci. 17: 657
196. Barshtein GR, Sibsai OYu et al. (1985) Dokl. AN SSSR 285: 371
197. Chan Y, White JH, Oyanagi Y (1978) Polym. Engng. Sci. 18: 268
198. Charrier JM, Rieger TM (1974) Fibre Sci. a. Technol. 7: 161
199. Malkin AYa (1975) Mekh. Polimerov 1: 173
200. Han CD (1976) Rheology in polymer processing. Academic, New York
201. Han CD (1974) J. Appl. Polym. Sci. 18: 2809
202. Tanaka H, White JL (1986) Polym. Engng. Sci. 20: 949
203. Faitelson LA, Yakobson EE (1977) Mekh. Polimerov 6: 1075
204. Faitelson LA (1978) Mekh. Polimerov 1: 113
205. Faitelson LA, Yakobson EE (1978) Mekh. Polimerov 1: 172
206. Vinogradov GV, Malkin AYa (1972) Intern. J. Polym. Mater. 1: 1
207. Ma ChY, White JL et al. (1985) Polym. Compos. 6: 215
208. Fridman ML, Popov VL et al. (1980) Dokl. AN SSSR 255: 1185
209. Fridman BL, Popov VP et al. (1982) Plast. Massy 2: 17
210. Pishin GA, Saveliev AP et al. (1988) Plast. Massy 3: 19
211. Lim TTS (1973) Polym. Engng. Sci. 13: 240
212. Tovamasyan YuM, Topolkaras VA et al. (1984) Plast. Massy 7: 33
213. Uriev NB (1980) Highly Concentrated dispersed systems Khimiya, Moscow
214. Guzeev VV, Rafikov MN, Malinsky YuM (1971) Plast. massy 12: 21
215. Kim VS, Skachkov VV (1988) Dispersion and blending as plastics manufacture processes, Khimiya, Moscow
216. Nielsen LE (1979) J. Polym. Sci.: Polym. Phys. 17: 1897
217. Nielsen LE (1966) J. Appl. Polym. Sci. 10: 97
218. Friedrich K, Karson UA (1983) Fibre Sci. Technol. 18: 37
219. Togin VA, Shchupak EN et al. (1984) Mekh. Kompoz. Mater. 4: 635
220. Shklovsky BI, Efres AL (1979) Alloyed semiconductor circuits, Nauka, Moscow
221. Togin VA, Timan SA et al. (1986) Dokl. AN SSSR 290: 154
222. Dzenis YuA (1986) Mekh. Kompoz. Mater. 1: 14
223. Hashin Z, Shtrikman S (1963) J. Mech. Phys. Solids 11: 127
224. Yanovsky YuG, Vinogradov GV, Barancheeva VV (1986) Mekh. Kompoz. Mater. 6: 107
225. Bartenev TN, Ermilova NE (1969) Kolloid. Zhurnal 31: 169
226. Pavlov VP, Vinogradov GV (1966) Kolloid. Zhurnal 28: 424

227. Onogi S, Matsumoto T, Warashina Y (1973) Trans. Soc. Rheol. 17: 175
228. Rosen SL, Rodrigues F (1965) J. Appl. Polym. Sci. 9: 1601
229. Vinogradov GV, Malkin AYa (1972) Int. J. Polym. Mater. 2: 1
230. Kambe H, Tokano M (1965) Proc. 4th Int. Congr. on Rheol. 4P3: 557 Interscience publ. New York
231. Zatsepina TI, Trapeznikov AA, Frolova EA (1969) Kolloid. Zhurnal 31: 207
232. Malkin AYa, Sabsai OYu et al. (1976) Kolloid. Zhurnal 38: 181
233. Fridman ML, Prut EV (1984) Usp. Khimii 53: 309
234. Zuev YuS (1979) Vysokomol. Soed. A21: 1203
235. Dolokova-Svehlova V (1982) J. Macromol. Sci. B21: 231
236. Zgaevsky NE, Ivan VV, Taranenko VK (1984) Mekh. Kompoz. Mater. 2: 233
237. Osnmyan VG, Knushyaku NM et al. (1984) Mekh. Kompoz. Mater. 3: 431
238. Brassell GW, Wischman KB (1974) J. Mater. Sci. 9: 307
239. Trachte K, Di Benedetto AT (1971) Int. J. Polym. Mater. 1: 75
240. Dekkers MEJ, Heikens D (1983) J. Appl. Polym. Sci. 23: 3809
241. Kardos JL (1984) Chem. Tech. 14: 431
242. Lee MCH (1987) J. Appl. Polym. Sci. 33: 2479
243. Vettegren VI, Dzyubenko PS et al. (1979) In: Structure and properties of polymeric materials. Zinatne, Riga
244. Monte SJ, Sugerman G (1981) Polym. Plast. Techn. a. Engng, 17: 95
245. Monte SJ, Sugerman S (1978) Plast. a. Rubb.: Mater. and Appl. 8: 117
246. Danusso F (1986) Polymer 27: 1385
247. Gorbatkina YuA (1987) Adhesion Strength in Polymer-Fiber Systems. Khimiya, Moscow
248. Zelensky ES (1984) Usp. Khimii 53: 290
249. Bashtanik BI, Yakimenka YG, Ovcharenko VG (1984) Vopr. Khimii i khim. Tekhn. 74: 67
250. Bunina LO, Kesler GE, Grigoriants IG et al. (1980) Dokl. AN SSSR 253: 902
251. Lhymn Ch (1985) J. Mater. Sci. Lett. 4: 1323
252. Kovriga VV, Kuznetsova IG et al. (1980) Mekh. Kompoz. Mater. 5: 222
253. Kutsenko MA, Kuznetsova IG, Kovriga VV (1984) Mekh. Kompoz. Mater. 2: 224
254. Kutsenko MA, Murasot AM et al. (1987) Mekh. Kompoz. Mater. 2: 343
255. Dekkers MEJ, Heikens D (1983) J. Mater. Sci. 28: 3281
256. Dekkers MEJ, Heikens D (1984) J. Mater. Sci. Lett. 3: 307
257. Pittolo M, Rurford RP (1984) J. Mater. Sci. 29: 3330
258. Tshoegl NW (1971) J. Polym. Sci. 32: 239
259. Goldman AYa, Kokotov YuV et al. (1987) Mekh. Kompoz. Metr. 3: 532
260. Mikhalsky AI (1984) In: Advances in science and technology chemistry and high molecular technology series, VINITI Moscow, p 151
261. Han CA, Sanford C, Yoo H (1978) Polym. Eng. Sci. 18: 849
262. Cope Donald E, Linert E (1980) Plast. Engng. 6: 37
263. Ess JW, Hornsby PR et al. (1984) Plast. a. Rubb. Proc. and Appl. 4: 7
264. Kovalevsky T, Cabinski R et al. (1982) Colloid. a. Polym. Sci. 260: 652
265. Badran BM, Galesky AM, Kryszewski M (1982) J. Appl. Polym. Sci. 27: 3669
266. Bigg DM (1982) Polym. Eng. and Sci. 22: 512
267. Han CD, Van Den TW, Shete P (1981) Polym. Eng. Sci. 21: 196
268. Plueddeman EP (1982) Silane coupling agents, Plenum, New York
269. Richard C, Hung K, Schreiber HP (1985) Polym. Compos. 6: 201
270. Morrei SH (1981) Plast. a. Rubb.: Proc. and Appl. 1: 179
271. Zverlin VG, Gladchenko AA, Sazonov VV (1985) Plast. massy 8: 45
272. Zhurkov SN (1980) Fiz. Tv. Tela 22: 3344
273. Zhurkov SN (1983) Fiz. Tv. Tela 25: 3119
274. Vettegren VI, Gmel A (1976) Vysokomol. Soed. B19: 52
275. Enikolopyan NS, Melkumov AN et al. (1987) Dokl. AN SSSR 296: 372
276. Goede J, Müller W (1976) Vysokomol. Soed. 18: 737
277. Vinogradov VM (1974) In: Structural plastics, Khimiya, Moscow, p 46

278. Brautman L, Agarwal B (1974) Polym. Eng. and Sci. 7: 581
279. Paul D, Newman S (1981) In: Polymer blends., Mir, Moscow, p 20
280. Atkins AG (1975) J. Mater. Sci. 5: 819
281. Vinogradov VM, Rubinchik BE, Il'ina LL (1987) Plast. massy 3: 60
282. Trostyanskaya EB, Shadgina ZM, Vinogradov VM (1981) Plast. massy 3: 16
283. Semikolenova NV, Zakharov VA (1982) Complex organometallic catalysts for olefin polymerization, IKhF AN SSSR Publication 9: 51
284. Howard EG, Zipsomb RD et al. (1981) Ind. Eng. Chem. Product Res. a. Development 20: 421
285. Russiyan LN, Belova VN et al. (1986) Complex organometallic catalysts for olefin polymerization, IKhF AN SSSR Publication, 10: 5
286. Pomogailo AD, Torosyan AA et al. (1980) Dokl. AN SSSR 252: 1148
287. Raspopov LN, Shaknova GE et al. (1982) In: Complex organometallic catalysts for olefin polymerization. IKhF AN SSSR Publication, 8: 31
288. Prokhorova TG, Mikhailov AI, Belov GP (1988) Plast. massy 3: 26
289. Zvereva YuA, Popov VA et al. (1980) Dokl. AN SSSR 252: 1174
290. Popov VA, Zvereva YuA, Grishin AN (1982) Vysokomol. Soed. A24: 1654
291. Aivazyan GB, Airapetyan SM (1986) Arm. Khim. Zhurnal 39: 508
292. Smirnov VV, Tkachenko LA et al. (1984) Dokl. AN SSSR 278: 927
293. Stalnova IO, Popov VL et al. (1982) Plast. massy 3: 15
294. Armer SS, Dorofeeva TV et al. (1985) Vysokomol. Soed. B27: 433
295. Enikolopyan NS, Pshechenkov PN, Grigorov LN (1983) Dokl. AN SSSR 289: 140
296. Semikolenova NV, Nesterov GA et al. (1985) Vysokomol. Soed. A27: 1998
297. Enikolopyan NS, Fridman ML et al. (1985) J. Appl. Polym. Sci. 32: 6107
298. Fridman ML, Brikenshtein Kh.-MA et al. (1986) Complex organometallic catalysts for olefin polymerization. IKhF AN SSSR Publication, 10: 72
299. Gevorgyan MA, Sivergin YuM et al. in: cf. No 298 9: 136
300. Nikitin LN (1982) Investigation of the structure and physical properties of Norplast and polyethylene/perlite based compositions, Candidate's thesis. VPI Voronezh
301. Novokshonova LA, Raspopov LN et al. (1975) Dokl. AN SSSR 224: 384
302. Belov GG, Diachkovsky FS et al. (1977) Plast. massy 10: 50
303. Aravin AG, Belov GG et al. (1986) In: cf. No 298, 9: 120
304. Aravin AG, Solomennik VD et al. in: cf. No 298, 9: 123
305. Chacko VP (1980) J. Polym. Sci.: Polym. Phys. Ed. 20: 2177
306. Abramova IM, Bunina LO et al. (1987) Plast. massy 6: 17
307. Skazka VS, Ruchin AE et al. (1984) Vysokomol. Soed. B26: 598
308. Lee BL, Nielsen LE (1977) J. Polym. Sci.: Polym. Phys. Ed. 15: 683
309. Grigorov LN, Shklyarov EI et al. (1983) Dokl. AN SSSR 270: 351
310. Enikolopyan NS, Berlin YuA et al. (1981) Dokl. AN SSSR 258: 1400
311. Berlin YuA, Boshenko SI et al. (1982) Dokl. AN SSSR 287; 1390
312. Grigorov LN, Shklyarova EI et al. (1984) Dokl. AN SSSR 278: 911
313. Yakova LL, Kuznetsov VA et al. (1977) Zh. Fiz. Khimii 51: 1451
314. Yakova LL, Kuznetsov VA, Tolstaya SN (1977) Dokl. AN SSSR 237: 1418
315. Yakova LL, Kuznetsov VA, Tolstaya SN (1984) Kolloid. Zhurnal 40: 590
316. Popov VA, Guzeev VV et al. (1984) Vysokomol. Soed. A26: 2489
317. Dmitrienko AV, Ivancheev SS et al. (1988) Vysokomol. Soed. A30: 72
318. Yushkova SM, Guzeev VV et al. (1982) Vysokomol. Soed. B24: 483
319. Tager AA, Yushkova SM (1984) In: Physical chemistry of polymer synthesis and processing, Univ. of Gorky press Gorky, p 52
320. Dubnikova IL, Meshkova IN et al. (1985) Dokl. AN SSSR 281; 629
321. Dubnikova IL, Meshkova IN et al. (1986) Complex organometallic catalysts for olefin polymerization, Publication IKhF 10: 27
322. Topolkaraev VA, Tovmasyan YuM et al. in: cf. Ref. No 321, 10: 100
323. Topolkaraev VA, Tovmasyan YuM et al. (1987) Mekh. Kompoz. Mater. 4: 616
324. Optov VA, Belov GP, Sabsai OYu in: cf. No 321, 10: 173
325. Sukhova TA, Belov GP et al. in: cf. No 321, 10: 159

326. Maksimova TV, Stalnova IO et al. (1986) Plast. massy 6: 50
327. Maksimova TV, Stalnova IO et al. in: cf. No 321, 10: 151
328. Telesnov VA, Zharov AI, Mironov NA (1985) Plast. massy 2: 25
329. Zvergin VG, Gladchenko AN, Sazonov WV (1984) In: Quality and efficiency control and design and fabrication of polymer processing equipment. VNIIRTmaxsh Publication, Tambov, p 32
330. Zvergin VG (1982) Khim. i Neft. Mashinostroyeniye 4: 20
331. Gevorgyan MA (1982) Investigation of polymerization-filled polyethylene composites, Institute of Chemical Physics Candidate's thesis
332. Frolova IN, Motavkin AV et al. (1987) Plast. massy 5: 15
333. Zaremsky NYu, Stepanyam AO et al. (1987) Vysokomol. Soed. A29: 242
334. Tager AA, Yushkova SM et al. (1987) Vysokomol. Soed. B29: 734
335. Olenin AV, Khristyuk AL et al. (1983) Vysokomol. Soed. A25: 423
336. Revjako MM, Markina AYa (1981) Khim i Khim. Tekhnol. (Minsk) 16:
337. Markina AYa, Tutaeva ML (1985) Khim. i Khim. Tekhnol. (Minsk) 20: 79
338. Paus KF, Illechev IE et al. (1986) Plast. massy 4: 18
339. Kataoka T, Kitano T (1979) Rheol. Acta 18: 635
340. Nicodemo L, Nikolais L (1974) J. Appl. Polym. Sci. 18: 2809
341. Kataoka T, Kitano T (1978) Rheol. Acta 17: 149
342. Yanovsky YuG, Vinogradov GV et al. (1985) Vysokomol. Soed. A27: 2385
343. Christensen RB (1969) J. Mech. Phys. Solids 17: 23
344. Dmitrienko AV, Demidova VA et al. (1985) Mekh. Kompoz. Mater. 3: 437
345. Kakhramanov NT, Abbasov AM et al. (1984) Mekh. Kompoz. Mater. 4: 707
346. Goldsmith HL, Mason SS (1967) In: Rheology theory and application. Academic, New York, p 86
347. Serge G, Silberberg A (1961) Nature 6476: 209
348. Karnis A, Goldsmith HL, Mason GS (1963) Nature 4902: 159
349. Bright PF, Crowson RJ, Folkes MJ (1978) J. Mater. Sci. 13: 2497
350. Ginffria R, Charhart RO, Davis DA (1963) J. Appl. Polym. Sci. 7: 1732
351. Cubat J, Szalanczi A (1974) Polym. Eng. Sci. 14: 873
352. Helger RP, Mennig G (1985) Polym. Eng. Sci. 25: 395
353. Wu S (1979) J. Rheol. 23: 415
354. Fakirov S, Fakirova C (1985) Polym. Compos. 1: 41
355. Sanon M, Chung B, Cohen C (1985) Polym. Eng. Sci. 25: 1008
356. Friedrich K (1981) Pract. Metalorg. 18: 513
357. Malrahu JC, Shulz JM (1986) Compos. Sci. a. Technol. 25: 187
358. Paus KF, Iliechev IE et al. (1986) Plast. massy 4: 18
359. Malkin AYa, Leonov AI (1970) In: Advances in polymer rheology, Khimiya, Moscow, p 98
360. Vinogradov GV (1974) In: Chemistry and technology of high molecular compounds. Advances in Science and Technology Series. VINITI Publication, Moscow, p 130
361. Schriedrum H (1979) Plastverarbeiter 3: 122
362. Baver B (1979) L'Officiel des plastiques et du Caoutchouk 269: 264
363. Korotchenko V, Bell T (1978) Heat Treatment of Metals 4: 88
364. (1978) Kunststoffe-Information 360: 223
365. Fridman ML (1981) Control of rheological properties of thermoplastics and compositions on their basis with a view to intensify the shaping process, Doctorate thesis, IKhF AN SSSR, Moscow
366. Faigelson LA, Yakobson EE (1987) Mekh. Kompoz. Mater. 1: 112
367. Bright PF, Darlington MW (1981) Plast. and Rubb. Proc. a. Appl. 1: 139
368. Motavkin AV, Martynov VM, Teleshov VA (1973) Mekh. Polimerov 3: 507
369. Motavkin AV, Kalinka YuA, Teleshev VA (1974) Mekh. Polimerov 1: 47
370. Motavkin AV, Ozola VYa, Krivatkin AM (1983) Plast. massy 6: 31
371. Turkevich R-von, Erwin L (1983) Polym. Eng. Sci. 23: 743
372. Tovmasyan YuM, Topolkaraev VA, Berlin AA (1986) Vysokomol. Soed. A6: 1162
373. Braznik YuV, Basntanik PI, Sporyagin EA (1985) Mekh. Kompoz. Mater. 5: 932

374. Busygin C, Martiner GM et al. (1983) Polym. Eng. Sci. 23: 766
375. Maine FW, Shepherd PD (1974) Composites 9: 193
376. Stalnova IO, Romanova ZA et al. (1984) Mekh. Kompoz. Mater. 4: 738
377. Lobkova ML, Stalnova IO et al. (1985) Plast. massy 2: 43
378. Zavodnikova LI, Popov VA et al. (1985) Plast. massy 1: 50
379. Paus KF, Ilieche IE et al. (1986) Plast. massy 1: 13
380. Sizova MD, Karmilova LV et al. (1986) Complex organometallic catalysts for olefin polymerization. IKhF AN SSSR publication 10: 93

Editor: N. S. Enikolopyan
Received December 18, 1989

Rheology of Filled Polymers

Alexander Y. Malkin
Research Institute of Plastics, Perovskii proezd 35, USSR 111112, Moscow E-112, USSR

Published data, concerning the results of measuring rheological properties of polymer melts with rigid fillers, are discussed. The review is based mainly on the works of the 1980s. The effect of a filler's presence on three fundamental parameters of a material, namely yield stress, viscosity, and rubbery elasticity is considered. The analysis is based essentially on qualitative considerations with respect to physical phenomena appearing when a rigid filler is introduced into the melt. General rules associated with the effect of concentration, intermolecular interaction, size and form of filler particles on rheological properties of a composition are traced. The review does not mention hydrodynamic phenomena, specific for flows of liquids with rigid filler, and also rheology of non-flowing multicomponent materials.

1 Introduction 70

2 Yield Stress 71
2.1 Definition 71
2.2 Methods 74
2.3 Temperature Dependence 77
2.4 Molecular Parameters 77
2.5 Concentration Dependence 80
2.6 Shape of Particles 81
2.7 Extension 81

3 Viscosity 82
3.1 General. Newtonian Flow 82
3.2 Non-Newtonian Flow 85
3.3 Wall Slip 87
3.4 Non-spherical Particles 88
3.5 Adsorption Effect 90
3.6 Extensional Flow 91

4 Elasticity 92
4.1 Normal Stresses 92
4.2 Dynamic Properties 93

5 Conclusion 95

6 References 96

Advances in Polymer Science 96

1 Introduction

First of all let us determine the genre of the present article. It is certainly not an exhaustive review since a review must contain references to hundreds of original works and the list of literature would occupy all the volume planned for the article. The paper is also not an original scientific report because it is based not only on my own data, but to a great extent on the discussion of experimental data of many authors. Most likely, it should be considered as a critical analysis in the form of an enlarged summary (or abstract) whose aim is to discuss the situation in the study of rheological properties of polymer melts, containing a rigid, filler, first of all bearing in mind the achivements of the last 5–7 years.

Analyzing the behavior of filled polymers, as any other heterophase systems, two aspects should be distinguished. First, these are the properties of such systems, i.e. their inherent characteristics, independent of a measuring method if, of course, the measurements are correct (to select criteria of "correctness" of an experiment, carried out with multiphase systems, seems to be an independent and by no means a simple problem). Second, this is a manifestation of these properties when heterophase systems flow in channels of different geometrical form. Behind all this stands the basic applied problem—finding out how the properties of filled polymers, appearing during their flow, affect the properties of finished articles.

Rheological properties of filled polymers are characterized by the same parameters as any fluid media [1]. These are:

apparent shear viscosity η; dependent on shear rate $\dot{\gamma}$ or shear stress τ;
elongational viscosity for uniaxial extension λ;
real G′ and imaginary G″ components of a complex dynamic modulus dependent on frequency ω;
yield stress Y;
first difference of normal stresses σ;
shear rubbery deformations γ_e and elastic (rubbery) modulus G_e.

All these quantities may depend on the content and behavior of a filler and to a certain extent on the prehistory of the material. This is a specific character of rheological properties of filled polymers.

Thus, estimating the properties of filled polymers, an investigator rests on a conventional system of simple indices, which are measured in unidimensional (desirably-viscometric) fields under the conditions of stationary flow or low periodic deformations. This, of course, does not exclude the possibility of going over to more complicated conditions of deformation, but as in other cases, it may turn out to be difficult to handle the results of such measurements.

On the basis of the results of rheological measurements the following problem is solved: formulation of a rheological equation of state, which gives grounds for a quantitative consideration of its behavior during the flow in channels of arbitrary form. However, it should be noted that there are hundreds of works discussing rheology of filled systems, while the next steps are made by very few. Therefore, the question whether it is possible to predict the behavior of filled polymers proceeding from their properties remains open, and its solution, as a rule, is restricted by qualitative considerations.

The manifestation of rheological properties of filled polymer melts is a combination of unusual (as compared to pure polymer melts) effects observed during their flow. These effects may turn out to be very diverse, since they depend on a geometrical scheme of deformation and the latter can vary in an arbitrary way. Therefore it is impossible to list them beforehand. Only one phenomenon must be mentioned because it must be regarded as a general phenomenon typical of the hydrodynamics of filled polymers. This is a wall effect, caused by a nonhomogeneous distribution of the concentration in a flow and formation near the wall of a layer depleted of a filler. Owing to this, an apparent "slippage" appears along a wall, a comparatively low-viscosity, layer formed by a dispersion liquid.

Analysis of rheological properties of filled polymers consists, in essence, of answers to a few general questions. They may be stated as follows:
In what way does the viscosity of dispersion depend on concentration of the solid phase?

What role do the dimensions and form of particles of the solid phase play?

How does loading affect rubbery elasticity of polymer melts?

What are the conditions for the appearance and the nature of concentrational dependence of the yield stress?

How do the viscoelastic properties of polymers change upon filling?

Some of these questions have strict and unambiguous answers, in a mathematical model, to other answers are derived from extensive empirical material. The present paper will discuss the problems formulated above, but concerning only rheological properties of filled polymer melts, leaving out the discussion of specific hydrodynamic effects occurring during their flow in channels of different geometrical form.

2 Yield Stress

2.1 Definition

The existence of yield stress Y at shear strains seems to be the most typical feature of rheological properties of highly filled polymers. A formal meaing of this term is quite obvious. It means that at stresses lower than Y the material behaves like a solid, i.e. it deforms only elastically, while at stresses higher than Y, like a liquid, i.e. it can flow. At a first approximation it may be assumed that the material is not deformed at all, if stresses are lower than Y. In this sense, filled polymers behave as visco-plastic media with a low-molecular and low-viscosity dispersion medium. This analogy is not random as will be stressed below when the values of the yield stress are compared for the systems with different dispersion media. The existence of yield stress in its physical meaning must be correlated with the strength of a structure formed by the interaction between the particles of a filler.

Though it is easy to determine formally what is a yield stress, in practice its measurement faces essential methodical and principal difficulties. Here two approaches are basically used, "stationary" and "dynamic". During stationary measurements a flow curve is measured and interpretation of the results obtained leads to the definition of a yield stress. An example of such an approach is given in Fig. 1, where experimental points are shown conventionally. They can be obtained under the con-

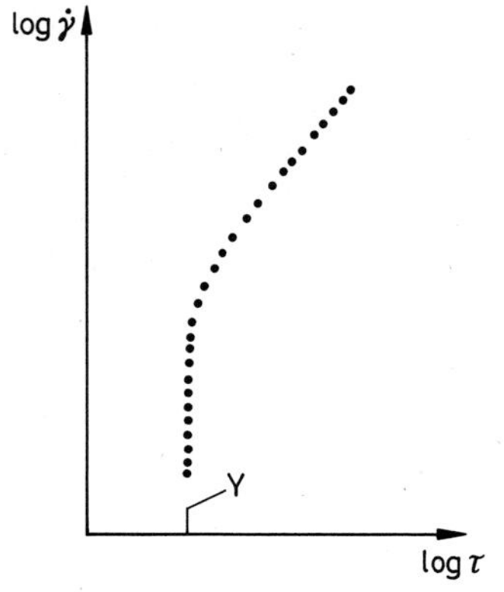

Fig. 1. A typical pattern for arrangement of experimental points during measurement of the flow curve for systems with yield stress

ditions $\dot{\gamma}$ = const or τ = const. This is insignificant since we speak about stationary measurements which are independent of time. Visually it can be seen that there is no flow to the left of the vertical row of points ($\dot{\gamma} \to 0$) and therefore the stress corresponding to this vertical must be taken as a yield stress.

However this is only a schematic (though quite realistic) representation. In reality there are following basic distinctions:

First. Points on the flow curve of a filled system may not coincide in the region of low stresses, if, during measurements, the stress is gradually increased or decreased. This means that in our case positions of points near the yield stress (and sometimes along the whole flow curve) may depend on the prehistory of the material's deformation, in particular, duration of rest, or time and intensity of deformation. This situation occures if a material possesses (and filled materials often do) thixotropic properties. Then during scanning from the side of low stresses, we will have to deal with the strength of an unfailed structure or the highest value of yield stress. In all the other cases the value of the yield stress being measured turns out to be lower because it corresponds to the strength of a structure partially failed or incompletely reconstructed.

Second. Constructing the flow curve (as in Fig. 1) we see that at $\tau < Y$ the flow is absent, no matter how long the stress may act. This is in conformity with the concepts of strength as a characteristic of a substance independent of time. Meanwhile, it is well known that in effect the stress at which the material fails depends on a period of action

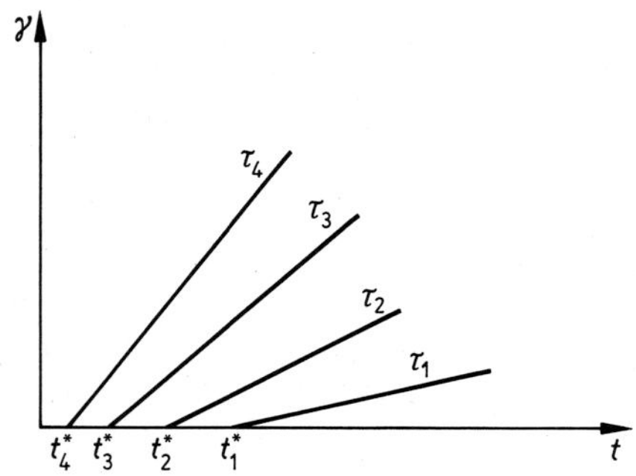

Fig. 2. The pattern of developing deformations in filled systems, in which durability of structure t_i^* depends on the applied stress τ_i. In the figure $\tau_4 > \tau_3 > \tau_2 > \tau_1$

of the load, i.e. strength properties of a substance are determined by its long-life properties. The same can be said about the strength of a structure formed by a filler in dispersion systems. Different situations are possible here. Thus, the following pattern of deformation development was observed in filled flowing media shown schematically in Fig. 2. When a certain stress τ_i is created, a material does not flow (is not deformed) at $t < t_i^*$, and then trivial plastic deformations are observed. It is shown in Fig. 2 that at $\tau_4 > \tau_3 > \tau_2 > \tau_1$ we have: $t_4^* < t_3^* < t_2^* < t_1^*$. This means that the values of time t_i^* may be interpreted as a durability of the structure under the action of stress τ_i, moreover, as the experiments have shown, with an increase in the durability decreases according to logarithmic law:

$$t^* = a - b \ln \tau \tag{1}$$

where a, b are empirical constants [2].

It should be noted that this is quite an unusual law, since in other known cases durability of solids is expressed by stronger laws, namely, exponential or power laws. Thus, in the given example we cannot give a unified definition of yield stress. The work cited is the only published observation of the durability of a filler's structure in dispersion systems. Therefore at present it is difficult to say how much such phenomena are typical for filled polymers, but we cannot exclude them.

Third. Points in Fig. 1 show only a part of a flow curve. In reality if we take very low stresses τ, it turns out that in this field a flow is also possible and in fact such â flow curve in the field of low rates looks like it is shown in Fig. 3, i.e. at low stresses τ the flow takes place, though viscosity in this range of stresses η_c turns out to be very high amounting to 10^9–10^{10} Pa s and, moreover, (and this is of fundamental importance) η_c in no way correlates with the viscosity of the dispersion medium.

Thus, measuring rheological properties of dispersions of low-molecular-weight (with viscosity about 60 Pa s) and high-molecular-weight poly(isobutilene) (with viscosity about 10^5 Pa s) with the same content of filler we see that the values of "creep viscosity" η_c of these systems are practically equal, in spite of the difference of the viscosity of the dispersion medium more than 1000 times [3].

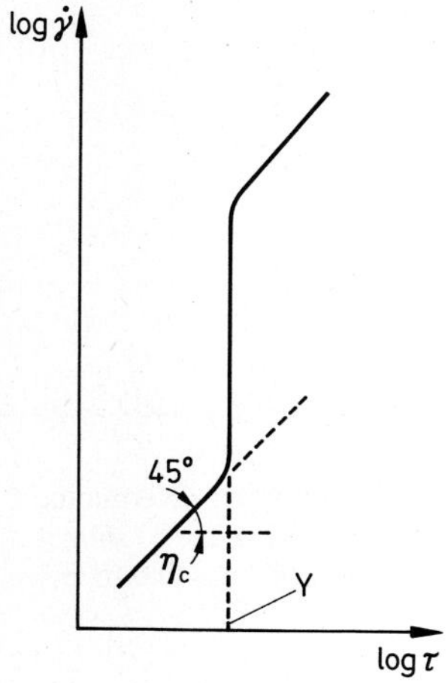

Fig. 3. A complete flow curve of plastic disperse system with a field of flow (with a very high viscosity η_c) at stresses smaller than the yield stress

This means that in the range of stresses which are lower than the apparent yield stress ($\tau < Y$), a mechanism of plastic flow ("creeping" flow) occurs differing completely from that in the common range.

The same situation, as stated in papers of G. V. Vinogradov's school in the 1950s, takes place during creep of plastic dispersion systems, i.e. low-molecular-weight liquids with netforming (active) fillers.

In practice, the phenomenon of creeping flow at $\tau < Y$ can usually be neglected. Thus, certainly, it is insignificant in the treatment of filled polymers, though it may be important, for example, in the discussion of the cold flow of filled elastomers. However, we cannot forget the existence of this effect, to say nothing of the particular interest of the physist in this phenomenon, which is probably similar to the mechanism of flow of plastic crystals.

2.2 Methods

Taking into account all the various stipulations, we shall still assume that yield stress has a certain physical meaning and it can be measured by a stationary method proceeding from the flow curve. However, to measure the points and achieve such a clear pattern as shown in Fig. 1 is not always convenient and it is rather labor-consuming. In practice, it is convenient to use a semi-analytical procedure. It is based on the utilization of an equation for flow curves taking into account the existence of yield stress. The most widespread equation of this kind is the Casson equation. It assumes that the $\tau(\dot{\gamma})$ dependence for filled polymers is expressed in the following way:

$$\tau^{1/2} = Y^{1/2} + a\dot{\gamma}^{1/2} \tag{2}$$

Then, plotting $\tau^{1/2}$ as a function of $\dot{\gamma}^{1/2}$ by a point of intersection of the straight line (provided a straight line is obtained) with the y-axis, we get the value of Y. Figure 4 shows conventionally two series of points for systems with a different content of the filler, for the straight line 2 the content of the filler is higher than for the line 1 and, accordingly, the value of Y is greater ($Y_2 > Y_1$).

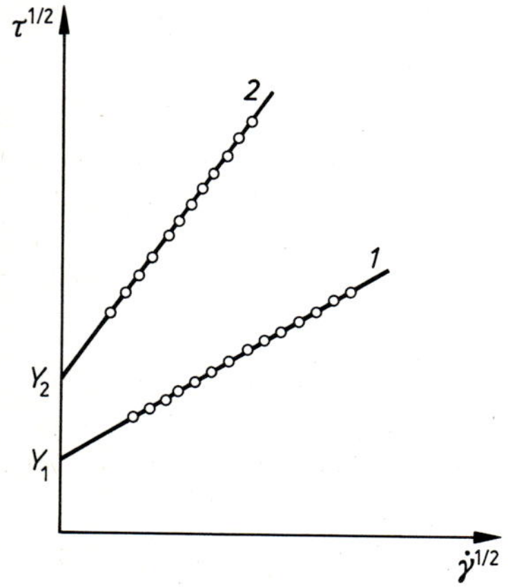

Fig. 4. Diagram illustrating the method for determining a yield stress by constructing a flow curve in the coordinates of the Casson equation. The content of the filler is more for curve *2* than for curve *1*

If in the coordinates $\tau^{1/2} - \dot{\gamma}^{1/2}$ the line obtained is not straight, we use a more general Schulman equation

$$\tau^{1/n} = Y^{1/n} + b\dot{\gamma}^{1/m} \tag{3}$$

where the values of n and m are selected empirically so as to straighten experimental points in the coordinates $\tau^{1/n} - \dot{\gamma}^{1/m}$.

It is only natural that computers present a powerful method for determining yield stress with the help of Eq. (3). A corresponding program of computations is based on such a selection of constants b, n, and m that provides a minimum deviation of the theoretical curve from experimental points.

During dynamic measurements frequency dependences of the components of a complex modulus G^* or dynamic viscosity $\eta'(\eta' = G''/\omega)$ are determined. Due to the existence of a well-known analogy between the functions $\tau(\dot{\gamma})$ or $G''(\omega)$ as well as between G' and normal stresses at shear flow σ, seemingly, we may expect that dynamic measurements in principle will give the same information as measurements of the flow curve [1].

Nevertheless, dynamic measurements have their own specific problems:

First. The problem of a limit of linearity has assumed a certain importance for investigating dynamic properties of filled polymers [4, 5]. Even for very small (from the point of view of measuring rheological properties of pure polymer melts) amplitudes of deformation, the values of a modulus depend on the amplitude.

The physics of this effects is quite understandable. Indeed, polymers by their nature are capable of great reversible deformation and therefore linearity of their mechanical behavior remains up to deformations of the order of 100%. But the structure formed by a filler undergoes brittle failure and hence, even for very small deformations the material changes and linearity of its behavior vanishes.

Here a complicated problem arises for finding a correct modulus, since it depends on the amplitude of deformation. There is no generally accepted method of extrapolating the dependence of the modulus on the amplitude of A to $A \to 0$. Therefore, here during the discussion of experimental data there may always be disagreements.

Second. The interpretation of the results of dynamic measurements is far from being obvious. Figure 5 gives a schematic representation of some typical results of measuring frequency dependence of the modulus $G'(\omega)$ for compositions with different concen-

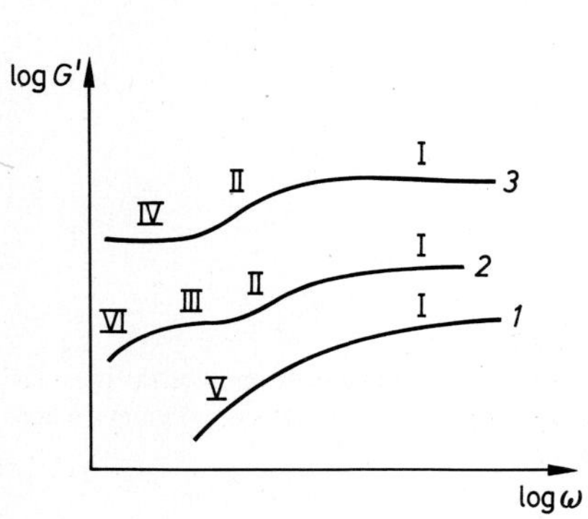

Fig. 5. Typical forms of frequency dependences of dynamic modulus. The content of the filler increases upon the transition from durve *1* to *2* and to *3*. The discussion of regions *I–VI*, displayed on the curves, see text

trations of the filler ($\varphi = 0$ or $\varphi \ll 1$ for curve 1, φ takes an intermediate value for curve 2 and is close to the limit for curve 3). A few typical fields are distinguished on the $G'(\omega)$ dependences. Let us discuss their physical meaning. Field I is the most clear, this is a rubbery plateau. Field V corresponds to the terminal zone (field of viscous flow). For curve 3 field IV may certainly be identified with yield stress, and field II with the transient zone of viscoelastic behavior. It is more complicated with curve 2 in Fig. 5. In the literature, field III is often identified with field IV of curve 3, i.e. with yield stress (for example [4]). It seems that there is an analogy between curve 2 in Fig. 5 and a "complete" flow curve with the field of flow at $\tau < Y$ in Fig. 3, the latter corresponds to field IV in Fig. 5.

However, two objections may be put forward against this interpretation: *First*, in many cases described, the curve of type 2 is observed for the systems in which a filler seemingly cannot form a "structure" possessing a certain strength, for example if such a filler is a glass sphere with a comparatively low concentration [4]. A possibility for forming such a structure is clear when an active filler is used, for example, carbon black, but is not at all obvious for rather large regular solid particles, because between them there is always a layer of polymer capable of flowing. *Second*, the range of frequencies during the transition from field III to IV corresponds to other (differing by several orders of magnitude) viscosity variations, than that which occurs at $\tau < Y$ in Fig. 3, so that a direct analogy is impossible here. Therefore, on the whole, curve 2 can hardly be treated as evidence of the existence of yield stress, but rather reflects the peculiarities of viscoelastic behavior of a polymer melt, modified by the presence of a solid filler. The latter leads to the appearance of new relaxation phenomena (for example in surface layers of a filler).

In order to complete the discussion of methodical problems, we should mention two more methods of determining yield stress. Figure 6 shows that for plastic disperse systems with low-molecular dispersion medium, when a constant rate of deformation, $\dot{\gamma} = \text{const.}$, is given, the dependence τ on time t passes through a maximum τ_m before a stationary value of shear stress τ_s is reached. We may assume that the value of the maximal shear stress τ_m is the maximum strength of the structure which must be destroyed so that the flow can occur. Here τ_m as well as τ_s do not depend or depend weakly on $\dot{\gamma}$, like Y. The difference between τ_m and τ_s takes into account the difference between maximum stress and yield stress. For filled polymer melts at low shear rates $\tau_m \approx \tau_s$, i.e. these quantities can be identified with Y.

There exists one more formally strict method for determining yield stress. It is based on the fact that during relaxation, the stress $\tau(t)$ cannot become lower that yield

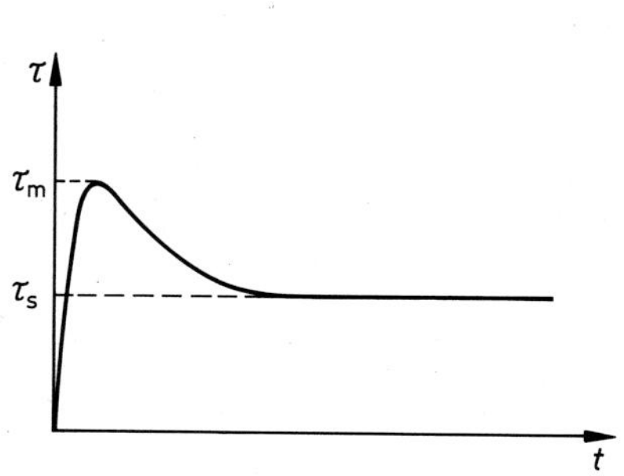

Fig. 6. A pattern of development of stresses in time $\tau(t)$ in a plastic disperse system (and filled polymer) when a constant shear rate is given

stress. Then when relaxation is measured after a steady-state flow has stopped, the material retains a residual stress, which, in its physical meaning is equivalent to Y. This method, in spite of the fact that it reflects a quite clear physical concept, is not used for filled polymer materials.

Now, on the basis of much experimental work, we shall answer the main questions concerning yield stress manifestation.

2.3 Temperature Dependence

Does yield stress depend on temperature? Probably, not, and flow curves constructed at different temperatures look as is shown in Fig. 7, where the arrow indicates the direction of temperature increase.

Such a pattern of flow curves has been observed in many cases and so it should be assumed a general rule that yield stress is independent of temperature.

Proceeding from the nature of yield stress as a characteristic of strength of the structure formed by a filler, a situation can be imagined where the character of intermolecular contacts will depend on temperature to a large extent, or at least, will sharply change at certain temperatures of transition. However, this is rather an exotic case, and although researchers have not observed it, it cannot be excluded.

2.4 Molecular Parameters of a Polymer

The problem of the influence of molecular parameters of a polymer (i.e. of an average molecular weight and molecular-weight distribution) on yield stress is related with the problem of the role of viscosity of a dispersion medium.

In essence, it follows from Fig. 7 that the viscosity of a dispersion medium does not affect yield stress, since with the variation in temperature (as in Fig. 7), the viscosity of a polymer melt changes by a hundred times, while yield stress remains unchanged.

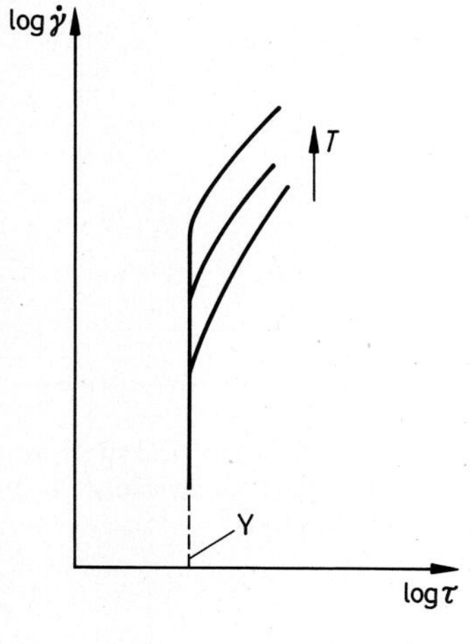

Fig. 7. A pattern for displacing flow curves of filled polymers with the growth in temperature. The *arrow* indicates the direction of temperature growth

A specific example of experimental data, quantitatively illustrating this conclusion, is represented in Fig. 8.

A comparison of values of yield stress for filled polymers of the same nature but of different molecular weights is of fundamental interest. An example of experimental results very clearly answering the question about the role of molecular weight is given in Fig. 9, where the concentration dependences of yield stress are presented for two filled poly(isobutilene)s with the viscosity differing by more than 10^3 times. As is seen, a difference between molecular weights and, as a result, a vast difference in the viscosity of a polymer, do not affect the values of yield stress.

It follows from Fig. 9, that in a polymer homologous sequence, yield stress in filled melts is independent of molecular weight.

It seems that results rather close in their meaning should be obtained, provided that monodisperse polymers are taken as dispersion media. Investigations into dynamic (viscoelastic) properties of such liquids with highly active filler have shown that [6],

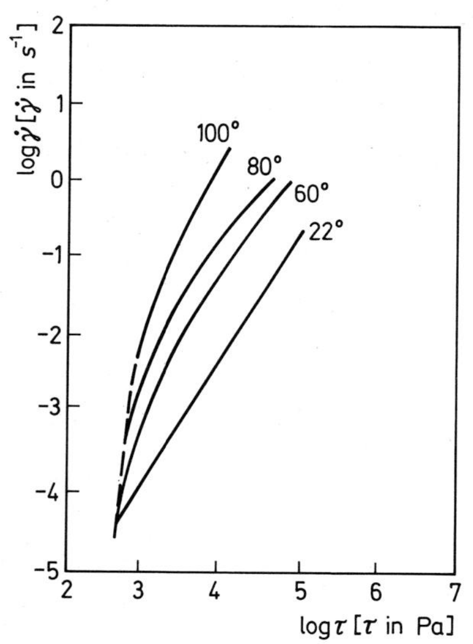

Fig. 8. Flow curves of dispersion of 5% acetylene carbon black (highly active) in poly(isobutylene) for different temperatures (indicated near curves)

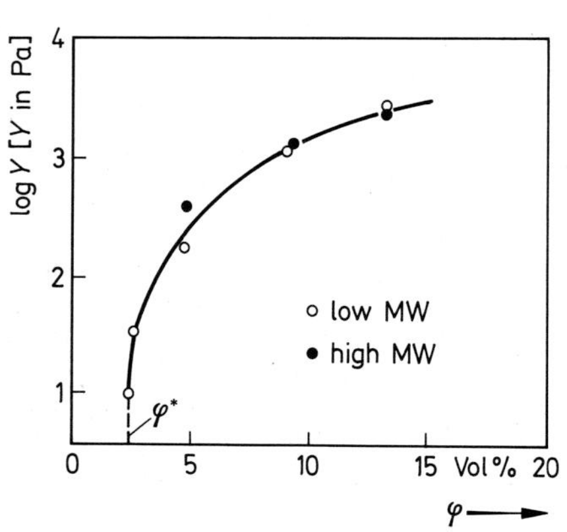

Fig. 9. Concentration dependences of yield stress for low- and high-molecular poly(isobutylene)s filled by highly active carbon black

in low-frequency field a plateau of the modulus (G′ as well as G″) appears, just like yield stress at stationary cases. The behavior of concentration dependence of the limiting value of the absolute value of modulus $|G_0^*|$ in this field is shown in Fig. 10. Moreover, this dependence corresponds to three different dispersion media, namely, monodisperse poly(butadiene) of different molecular weight and low-molecular silicone oil. The generality of the obtained $|G_0^*|$ vs φ dependence for different liquids indicates that it reflects only a net-formation ability of the filler and, as it is shown in Figs 8 and 9, it is not affected by the viscosity of a dispersion medium.

The problem of concentration dependence of yield stress will be discussed in detail below. Here we only note that (as is shown in Figs 9 and 10) yield stress may change by a few decimal orders while elastic modulus changes only by several in the field of rubbery plateau and, moreover, mainly in the range of high concentrations of a filler.

Data of Figs 8–10 give a simple pattern of yield stress being independent of the viscosity of monodisperse polymers, indicating that yield stress is determined only by the structure of a filler. However, it turned out that if we go over from mono- to polydisperse polymers of one row, yield stress estimated by a flow curve, changes by tens of times [7]. This result is quite unexpected and can be explained only presumably by some qualitative considerations. Since in case of both mono- and polydisperse polymers yield stress is independent of viscosity, probably, the decisive role is played by more fine effects. Here, possibly, the same qualitative differences of relaxation properties of mono- and polydisperse polymers, which are known as regards their viscosity properties [1].

According to the concepts, given in the paper [7], a significant difference between the values of yield stress of equiconcentrated dispersions of mono- and polydisperse polymers and the effect of molecular weight of monodisperse polymers on the value of yield stress is connected with the specific adsorption on the surface of filler particles of shorter molecules, so that for polydisperse polymers (irrespective of their average molecular weight) this is the layer of the same molecules. At the same time, upon a transition to a number of monodisperse polymers, properties of the adsorption layer become different.

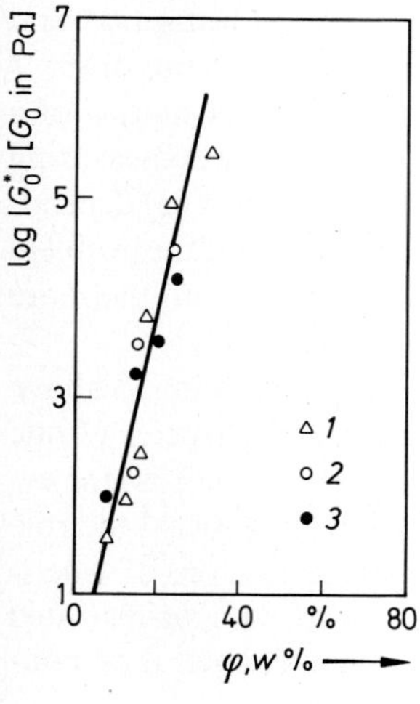

Fig. 10. Concentration dependence of a modulus in the region of low-frequency plateau (i.e. "yield stress", measured by a dynamic modulus). Dispersion medium: poly (butadiene) with M = 1.35×10^5 (*1*), silicone oil (*2*); polybutadiene with M = 1×10^4 (*3*). The points are taken from Ref. [6]. The *straight line* through these points is drawn by the author of the present paper. In the original work the points are connected by a curve in another manner

The role of interaction between a polymer and a filler in the net-formation is clearly manifested in the influence of a specific surface of a filler on the value of yield stress. As follows from qualitative considerations, with an increase in the surface of a filler the values grow, moreover, the variation may be 10-fold [7].

The interaction effect between a dispersion medium and a filler on the net-formation in a filled polymer melt is mostly visuallized in the fact that the finishing treatment of the surface of solid particles results in a significant change of the position of the flow curve of the filled polymer on the whole, and yield stress as well [5, 8–10].

The same is indicated by yield stress values (as the position of the whole flow curve) when we compare the properties of different polymer matrices with the same filler [11].

There is a great number of data indicating that the character of a net-formation in a disperse system both depends on the nature of a filler and dispersion medium.

Though such data are ambiguous, and sometimes even contradictory [12], they can be rationally explained on the basis of qualitative considerations on intermolecular interactions of a polymer with a filler. Of practical importance is the fact that varying the nature of the dispersion medium and the filler and thus controlling the intensity of net-formation, we can vary the yield stress of filled polymers within wide limits and in different directions.

The situation becomes most complicated in multicomponent systems, for example, if we speak about filling of plasticized polymers and solutions. The viscosity of a dispersion medium may vary here due to different reasons, namely a change in the nature of the solvent, concentration of the solution, molecular weight of the polymer. Naturally, here the interaction between the liquid and the filler changes, for one, a distinct adsorption layer, which modifies the surface and hence the activity (net-formation ability) of the filler, arises. Therefore in such multicomponent systems in the general case we can hardly expect universal values of yield stress, depending only on the concentration of the filler. Experimental data also confirm this conclusion [13].

2.5 Concentration Dependence

How does yield stress depend on a filler concentration? It is shown in Fig. 9 that appreciable values of Y appear beginning from a certain critical concentration φ^* and then increase rather sharply. Though the existence of φ^* seems to be quite obvious from the view point of the possibility of contacts of the filler, i.e. the beginning of a netformation in the system, practically the problem turns on the accuracy of measuring small stresses in high-viscosity media. It is quite possible to represent the $Y(\varphi)$ dependence by exponential law, as follows from Fig. 10, for example, leaving aside the problem of the behavior of this function at very low concentrations of the filler, all the more the small values of are measured with a significant part of uncertainty.

How does yield stress depend on the size of particles? We have mentioned above that increasing the specific surface, i.e. decreasing an average size of particles of one type, causes an increase in yield stress. This fact was observed in many works (for example [14–16]). Clear model experiments the purpose of which was to reveal the role of a particle's size were carried out in work [8]. By an example of suspensions of spherical particles in polystyrene melt it was shown that yield stress of equiconcentrated dispersions may change by a hundred of times according to the diameter d of nonagglomerating spheres, and here $Y \sim d^{-1}$, which is in agreement with some well-known theoretical estimates.

2.6 Shape of Particles

It follows from general considerations that the role of the shape of the filler particles during net-formation must be very significant. Thus, it is well-known that the transition from spherical particles to rod-like ones in homogeneous systems results in such radical structural effect as the formation of liquid-crystal phase. Something like that must be observed in disperse systems.

Figure 11 shows a qualitative form of the variation nature of flow curves upon introducing into the melt flexible (fibre-like) particles of an extended form. The flow curve of a dispersion medium in Fig. 11 is represented by a straight line, i.e. the medium is a Newtonian liquid. Introducing even small amounts of fibre-like particles results in the appearance of a sharply defined branch, but in the region of low stresses the possibility of the flow with a constant viscosity is retained. A sharp inflection on the flow curve (especially on the right-hand curve in Fig. 11) is to a certain extent analogous to yield stress, while a low-rate curve is analogous to the region of creeping flow in Fig. 3. However, this analogy is of a formal character, since when dispersion with anisotropic filler is deformed, a great irreversible effect appears: orientation of fibre-like particles, and macrostructure of a felt type is formed. Therefore the results of measurement determine the behavior rather than the properties of a filled melt.

Generally speaking, to obtain, reliable rheological characteristics of disperse systems with fibre-like filler turned out to be a difficult methodological problem. Therefore, the question on the effect of the shape of a filler particles on the value of yield stress is left open at present. In the papers published we can encounter only individual examples and qualitative considerations concerning this question, which do not enable us to formulate general conclusions.

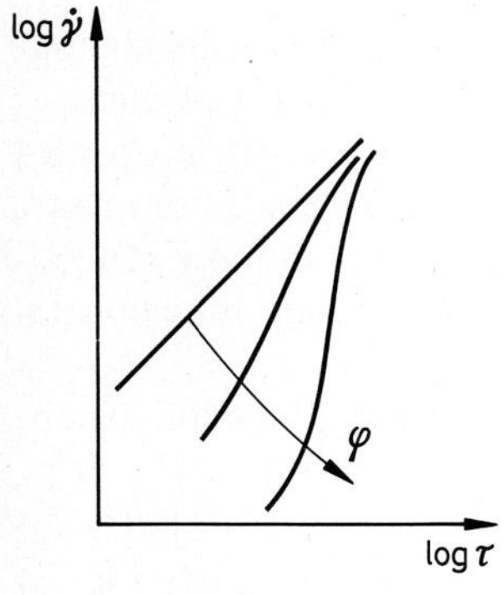

Fig. 11. Typical flow curves of dispersions of fibre-forming particles in a Newtonian liquid. The *arrow* indicates the direction in which the filler concentration increases

2.7 Extension

There are known very few investigations of the behavior of filled polymer melts at different stress states, distinct from shear, for example at uniaxial extension. One can draw from general considerations that since yield stress by its physical meaning corresponds to the limiting state (failure) of a solid structure, between yield stress, measured at a shear Y and a uniaxial extension Y_e. There should be correspondence as such a correspondence is set up between the conditions of failure of a solid at different geometrical schemes of its loading.

This approach was realized during the discussion of experimental data concerning the conditions of reaching yield stress, in papers [8, 17, 18]. In these works yield stress at the shear Y and at the uniaxial extension Y_e were found (with the help of Casson's equation) for one and the same objects. The results obtained by the authors have shown that $Y_e/Y \approx 1.4$–1.9, which was treated as a confirmation of von Mises criterion, which determines a general condition for failure of solids at different stressed states. According to this criterion the indicated ratio must be equal to $\sqrt{3}$, i.e. ~ 1.7. The problem on the ratio Y_e/Y essentially leads to a general problem of constructing a rheological equation of state for filled polymers.

It seems however that this problem is not fully cleared up. Thus, it was stated in paper [19] on the basis of experimental data obtained earlier that Y_e increased with a filler's concentration in proportion to $\varphi^{4.7}$. Such a law for the concentration dependence of yield stress at the shear $Y(\varphi)$ is not described. It is possible that since the behavior of the dependences $Y(\varphi)$ and $Y_e(\varphi)$ is different, there is no universal value of the Y_e/Y ratio. It is quite probable however, that indicated discrepancies follow just from different ways of analytical approximation of particular experimental data. The only unquestionable fact is that Y_e as well as Y grow very sharply with an increase in concentration.

3 Viscosity

3.1 General. Newtonian Flow

We can suppose the following diagram that shows how the system becoming complicated due to filling, changes its rheological and viscosity properties in particular.

This diagram essentially reflects the main features of variation of viscosity properties of a liquid when solid particles of a filler is introduced into it. It remains to impart a quantitative form to these features. This form is based on a great number of theoretical and experimental papers, which made it possible to formulate fundamental concepts in this field.

In the general case the dispersion viscosity (i.e. a filled system) as a function of concentration $\eta(\varphi)$ can be presented as a power series:

$$\eta = \sum_{n=0} a_n \varphi^n \tag{4}$$

Such representation of the $\eta(\varphi)$ dependence was discussed in dozens of papers. This enables us to confine ourselves to a short summary. The first term of the series a_0 (for $\varphi = 0$) obviously equals the viscosity of the dispersion medium:

Diagram of the Effect of a Filler on Viscosity Properties of Disperse (Filled) Systems

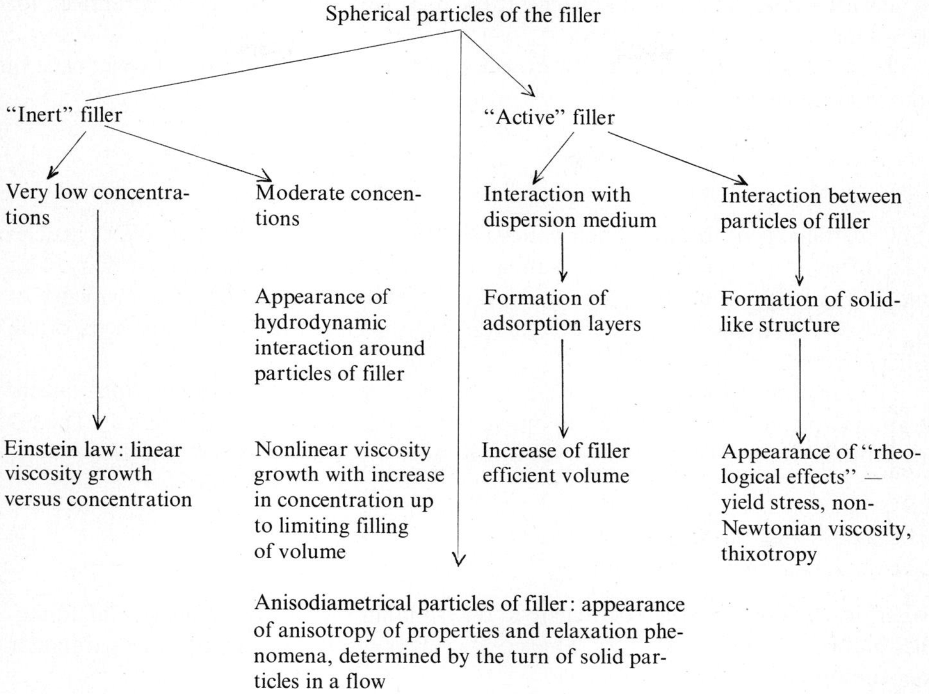

$a_0 = \eta_0$. Then dependence (4) can be conveniently presented in a dimensionless form — as specific viscosity η_{sp} vs dependence:

$$\eta_{sp} = \frac{\eta - \eta_0}{\eta_0} = \sum_{n=1} b_n \varphi^n \tag{5}$$

The value of the first coefficient b_1 for the dispersion of spherical particles is well known and generally accepted. This is Einstein coefficient $b_1 = 2.5$, taking into account the viscosity variation of the dispersion medium upon introducing non-interacting solid particles of spherical form into it. Thus, for $\varphi \ll 1$ $\eta_{sp} = 2.5\varphi$, i.e. the viscosity grows linearly with an increase in concentration, irrespective of a size of particles.

But even the value of the second coefficient b_2 is less determined. In different theoretical papers the values from 4.4 to 14.1 are encountered, though, probably, the most reliable value for spherical particles was given by Gut — Gold: $b_2 = 14.1$. As regards the higher terms of the series, they are still less determined.

The $\eta(\varphi)$ dependence in the form of Eq. (5) makes it possible to describe any experimental data by selecting coefficients b_i. But, nevertheless, formula (5) should be modified introducing a concept of "maximum degree of loading" φ_m. This is due to the fact that the values of φ cannot reach 1, since very large values of φ are devoid of physical sense: in a volume there cannot be more solid particles than a certain maximum amount φ_m. Therefore it is reasonable to normalize φ according to φ_m and consider η_{sp} as a function of reduced concentration $\varphi_r = \varphi/\varphi_m$. Obviously, by definition, for $\varphi_r \to 1, \eta_r \to \infty$.

In coordinates $\eta_{sp} - \varphi_r$ we can expect a generalized concentration dependence of the viscosity in the form:

$$\eta_{sp} = \sum_{n=1} c_n \varphi_r^n \tag{6}$$

The coefficients of this dependence, however, must not be equal for the filler particles of a different form; they grow with an increase in the degree of anisodiametricity. This means that the role of the higher terms (i.e. deviation from the linear dependence of η_{sp} on φ_r) is manifested at the lesser values of φ_r the more the anisodiametricity of the filler particles.

The function $\eta_{sp}(\varphi_r)$ of power type is not always convenient for presenting concentration dependence of the viscosity due to very many arbitrary coefficients c_n. The exponential Mooney formula deserves much more attention in this connection:

$$\eta_{sp} = \exp\left(1 + \frac{2.5\varphi}{1 - \varphi_r}\right) \tag{7}$$

where $\varphi_r = K\varphi$, and the coefficient K determining the maximum degree of filling, lies within the limits 1.35–1.91. This coefficient obviously is the only free parameter depending on the system structure.

For $\varphi_r \ll 1$ Mooney formula transforms in a natural way into Einstein equation, and for $\varphi_r \to 1$ this formula predicts that $\eta \to \infty$, which corresponds to the physical meaning of the phenomenon.

In special literature many other theoretical or empirical equations were suggested for describing η_{sp} vs φ dependence. Thus Thomas suggested [20] and Metzner cited [21] [21] the following empirical equation as successful:

$$\eta_{sp} = 2.5\varphi + 10.05\varphi^2 + A\exp(B\varphi) \tag{8}$$

where $A = 0.00273$, $B = 16.6$. This equation, according to the data of work [20] is fulfilled for suspensions of spherical particles in a fantastically wide range of values of particles' diameter (from 0.1 to 440 μ).

In a certain number of papers [22, 23] the authors successfully used the equation of the form:

$$\eta/\eta_0 = \frac{1}{(1 - \varphi/0.680)^2} \tag{9}$$

As the structure of Eq. (8) shows, the first coefficient for $\varphi \to 0$ in the power expansion of this equation is equal to Einstein coefficient (according to formula (9) it is somewhat higher, but not much).

To what extent are these or other equations, which were suggested for the description of the η versus φ dependence general or universal?

Of course, any set of experimental data can be described by selecting an appropriate empirical equation with an arbitrary set of constants. However, comparing a vast wealth of the known results of measurements of suspension viscosity, it should be admitted that a universal formula for the $\eta(\varphi)$ dependence does not exist, and significant discrepancies may begin already from a linear term, so that physical reasons for exagerated values of the coefficient b_1 as compared to 2.5 should be looked for.

As regards general principles, the main effects are manifested in the region of comparatively high degrees of filling, it does not matter if the case at hand is a hydrodynamic interaction of the flows around them or a direct intermolecular interaction of the filler's particles. Here we should bear in mind the following decisive principles.

First, it is necessary to take into account a yield stress, which at once makes the problem on "universality" of the $\eta(\varphi)$ dependence to be doubtful since it is unclear what viscosity is spoken about.

Second, if we speak about polymer melts, we should take into account their non-Newtonian properties, so that the $\eta(\varphi)$ dependence will be different at different rates or shear stresses. This, besides all other things, raises the question on the correct selection of conditions for comparing the viscosity of the systems with different filler content.

Third, a complicated question on the role of the dispersion of particles dimensions of particles dimensions is of independent value: it is known that the viscosity of equiconcentrated dispersions of even spherical particles depends on the fact if spheres of one dimension or mixtures of different fractions were used in the experiments and here in all the cases the transition from monodisperse particles to wide distributions leads to a considerable decrease in viscosity [21] (which, certainly, is of theoretical and enormous practical interest as well).

At least, in absolute majority of cases, where the concentration dependence of viscosity is discussed, the case at hand is a shear flow. At the same time, it is by no means obvious (to be more exact the reverse is valid) that the values of the viscosity of dispersions determined during shear, will correlate with the values of the viscosity measured at other types of stressed state, for example at extension. Then a concept on the viscosity of suspensions (except ultimately diluted) loses its unambiguousness, and correspondingly the coefficients c_n cease to be characteristics of the system, because they become dependent on the type of flow.

3.2 Non-Newtonian Flow

Since non-Newtonian flow is typical for polymer melts, the discussion of a filler's role must explicitly take into account this fundamental fact. Here, spoken above, the total flow curve includes the field of yield stress (the field of creeping flow at $\tau < Y$ may not be taken into account in the majority of applications). Therefore the total equation for the dependence of efficient viscosity on concentration must take into account the indicated effects.

There are possible different forms of analytic representation of the η versus $\dot{\gamma}$ dependence with an allowance made for non-Newtonian effects. To expose the decisive factors of this dependence an approach suggested in [19] deserves much attention. In this paper the following equation which is valid in the field of the flow, but taking into account also the presence of yield stress, is written:

$$\eta(\varphi) = \frac{Y}{\dot{\gamma}} + \eta_r(1 + \lambda\dot{\gamma})^{-n_1} \tag{10}$$

Concentration of the filler affects the constants of this equation Y, η_r and λ.

According to the structure of this equation the quantity φ indicates the influence of the filler on yield stress, and η_r on Newtonian (more exactly, "quasi-Newtonian" due to yield stress) viscosity. Both these dependences $Y(\varphi)$ and $\eta_r(\varphi)$ were discussed above. Non-Newtonian behavior of the dispersion medium in (10) is reflected through characteristic time of relaxation λ, i.e. in the absence of a filler the flow curve of a melt is described by the formula:

$$\eta = \eta_0(1 + \lambda_0\dot{\gamma})^{-n_1} \tag{11}$$

where η_0 is the initial Newtonian viscosity and λ the characteristic time of relaxation of a polymer melt.

Though the accuracy of description of flow curves of real polymer melts, attained by means of Eq. (10), is not always sufficient, but doubtless the equation of such a structure based on the idea of "relaxation" mechanism of non-Newtonian polymer flow, correctly reflects the main peculiarities of viscous properties. Therefore while discussing the effect a filler has on the viscosity properties of polymer melts, besides the dependences $Y(\varphi)$ and $\eta_r(\varphi)$ discussed above, it is important to establish the mechanism of the $\lambda(\varphi)$ dependence, since the filler modifies the characteristic time of relaxation. According to [19], a possible form of the λ versus φ dependence may be an expression

$$\lambda(\varphi) = \lambda_0(1 - \varphi)^{-n_3} \tag{12}$$

where n_3 is the empirical constant.

Concrete calculations carried out via formula (10) for different values of constant have shown that it reflects the behavior of flow curves quite really. However, a series doubt remains: for such a system (with a yield stress) it is not obvious how to determine the initial Newtonian viscosity (is it necessary to determine it and does it exist?).

The peculiarities of a filler's nature and liquid medium are manifested via the values of the constants of a rheological equation. They may be most various, but in all cases the main differences of flow curves have the greatest effect near the yield stress. Just here the flow curves diverge the most sharply, while at high rates the role of the content of the filler is levelled and the difference between flow curves of compositions with different content of the filler becomes insignificant. This is illustrated in Fig. 12, where region A corresponds conventionally to rates and stresses far from yield stress, and region B includes directly the yield stress. The flow curves of compsitions with a fibrous filler are arranged in the same way (Fig. 11).

Therefore in many cases, where the behavior of polymer compositions under real technological conditions is of practical interest, i.e. in the region of high deformation rates, the yield stress and the region near it may be neglected and only upper branches must be considered. There the influence of the filler on the viscosity of the melt turn out to be unexpectedly weak, at least incomparable weaker than at low shear rates. This becomes the most visual if the regions A and B in Fig. 12 are depicted separately: in the mutual position of the flow curves in these comparable cases there is very little in common.

A similar thing takes place when we consider flow curves obtained at different temperatures. As seen from Fig. 7, if we take a region of low shear rates, then due to the absence of the temperature dependence Y, the apparent activation energy vanishes. At sufficiently high shear rates, when a polymer dispersion medium flows, the activation energy becomes equal to the activation energy of the viscous flow of a polymer melt and the presence of the filler in this ratio is of little importance.

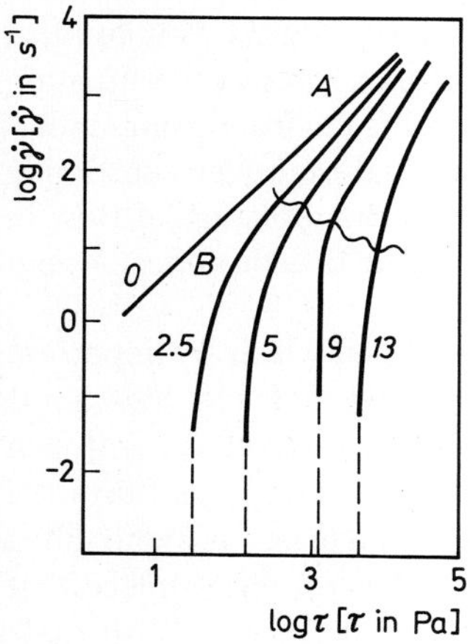

Fig. 12. Flow curves of poly(isobutylene), containing different concentrations of active filler (acetylene carbon black). Concentration (in volume percent) is indicated near the curves. *A* is the region of flow for stresses exceeding the yield stress; *B* is the region directly adjacent to the yield stress

3.3 Wall Slip

The fact that the appearance of a wall slip at sufficiently high shear rates is a property inwardly inherent in filled polymers or an external manifestation of these properties may be discussed, but obviously, the role of this effect during the flow of compositions with a disperse filler is great. The wall slip, beginning in the region of high shear rates, was marked many times as the effect that must be taken into account in the analysis of rheological properties of filled polymer melts [24, 25], and the appearance of a slip is initiated in the entry (transitional) zone of the channel [26]. It is quite possible that in reality not a true wall slip takes place, but the formation of a low-viscosity wall layer depleted of a filler. This is most characteristic for the systems with low-viscosity binders. From the point of view of hydrodynamics, an exact mechanism of motion of a material near the wall is immaterial, since in any case it appears as a wall slip.

In this sense for calculating a bulk output Q during the flow with wall slip as a first approximation the following formula can be used:

$$Q = Q_O + \pi R^2 V_s \qquad (13)$$

where Q_O is the bulk output determined by the flow in channel's volume, V_s is the velocity of wall slip, R is the radius of the channel.

The dependence of V_s on rheological parameters-shear stress on the wall and *fri*ction coefficient — as far as the author knows, for filled polymers was not investigated somewhat completely, though its determination is necessary for a specific solution of hydrodynamic problems related to the flow of filled polymers.

If a slip is in fact caused by the formation of a wall low-viscosity layer, depleted or completely devoid of a filler (i.e. if a layer of a comparatively low-viscosity dispersion medium was formed near a wall), this leads to a very general problem on radial migration of filler's particles and to the appearance of distribution of filler's concentration over the radius of the channel or, in more general case, in the direction perpendicular to the channel's axis. This problem was investigated theoretically for dilute (noninteracting) suspensions in a number of works. However, this theoretical scheme is very far from the real situation, which takes place during the flow of filled polymers in channels of different geometrical form. Therefore at present the situation with the distribution of a filler's concentration in a channel is far from being understood, and the information at hand is bounded by individual examples concerning a change in distribution of concentration along the across the direction of flow in end products [27], and the experimental data obtained indicate that this effect is very significant and affects the properties of the end product.

At high shear rates (or high flow velocities) other effects appear, which are important for understanding the behavior of highly filled polymers. They include rapid mechanic destruction of both dispersion medium, i.e. destruction of macromolecules [24], and to even greater extent destruction of the filler, particularly if long glass fibers are used as a filler [24, 28]. In addition, at high flow velocities of highviscous highly filled polymers an intensive dissipation of the work of viscous friction takes place that causes an increase in temperature of the material by tens of degrees [29]. All these effects are, of course, very essential for discussing the behavior of highly filled polymers during flow, but nevertheless they are beyond the scope of their rheological properties.

3.4 Nonspherical Particles

The representation of the dispersion viscosity given above related to suspensions of particles of spherical form. Upon a transition to anisodiametrical particles a number of new effects arises. The effect of nonsphericity is often discussed on the example of model dispersions of particles of ellipsoidal form. An exact form of particles to a first approximation is not very significant, the degree of anisodiametricity is only important, or for ellipsoidal particles, their eccentricity.

Even if the peculiarities of net-formation of nonspherical particles are not taken into account, at least two fundamentally new effects arise during the flow of dispersion. First, this is the possibility to be oriented in the flow, as a consequence of which the medium becomes anisotropic. And second, this is the possibility to rotate the spherical particles in the flow (spherical particles can, of course, rotato too, but their rotation does not affect the structure of the system as a whole).

If we consider a shear flow of a diluted suspension of noninteracting particles, then substitution of spheres by particles of ellipsoidal form leads only to a variation of

Einstein coefficient b_1 in (5) for viscosity 2.5 by a value dependent on the ratio between the lengths of the axes of ellipsoids. However, for the flows of different geometry (for example, uniaxial extension) the situation is rather complicated. Due to different orientation of ellipsoids upon shear and other geometrical schemes of flow, the correspondence between the viscosity changed at shear and behavior of dispersions at stressed states of other types is completely lost. Indeed, due to anisotropy of dispersion properties of anisodiametrical particles, the viscosity ceases to be a scalar property of the material and must be treated as a tensor quantity.

As regards shear flow the role of particle's geometry of the filler is rather unambiguous: with an increase in the ratio between the length of the filler and their diameter the viscosity of equiconcentrated dispersions increases [22], as it is schematically shown in Fig. 13. For the region $\varphi \rightarrow 0$ this effect results in an increase in the initial angle of slope of tangents to the dependences $\eta_{sp}(\varphi)$ as compared to $b_1 = 2.5$.

Quite specific effects in the flow of dispersions of long fibers are connected with particles orientation in the flow. Indeed, the state of fibers during the flow changes greatly as compared the initial state, so that the material in a steady-state flow is an anisotropic medium. Therefore the viscosity of such a suspension may become independent of a fiber's length [30]. The most strong effects caused by a deformation of anisotropic particles should be expected in transient flows, in particular if the particles themselves are flexible and deformed in the flow.

Here, of course, we may expect also an appearance of thixotropic phenomena, because a transition from the structure changed by deformation to a conventionally "homogeneous" structure of a filled system occurs rather slowly. And if an investigator at different moments of time deals with different structure of the medium, its properties will be different.

This is obvious for the simplest case of nondeformable anisotropic particles. Even if such particles do not change the form, i.e. they are rigid, a new in principle effect in comparison to spherical particles, is their turn upon the flow of dispersion. For suspensions of anisodiametrical particles we can introduce a new characteristic time parameter D_r^{-1}, equal to an inverse value of the coefficient of rotational diffusion and, correspondingly, a dimensionless parameter $C = \dot{\gamma} D_r^{-1}$. The value of D_r is expressed via the ratio of semiaxes of ellipsoid to the viscosity of a dispersion medium.

The existence of characteristic time D_r^{-1} must lead to the appearance of specific relaxation effects. This relaxation mechanism has nothing in common with visco-

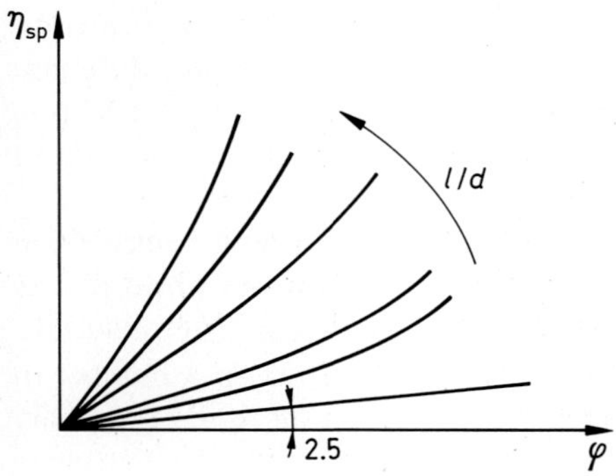

Fig. 13. Pattern of variation of concentration dependence of suspension viscosity when the ratio between the length and diameter of anisodiametricity of filler's particles increases. The *arrow* indicates the direction of growth of l/d of filler's particles. The slop of the initial part of line A (for spherical particles) is 2.5

elasticity of the medium. The indicated relaxation effect by its nature can be treated as a consequence of the transition (proceeding in time) from a preferred orientation of a filler particles, determined by the action of mechanical field, to statistically equiprobable distribution of orientations; the characteristic time of the process is D_r^{-1}. This new ("geometrical") relaxation mechanism, in principle, may lead to two important phenomena: first, to a non-Newtonian flow, and second, to the appearance of normal stresses during shear flow, which (this should be emphasized again) are not connected to the elasticity of the medium.

Indeed, as discussed above (Fig. 11), introducing into a Newtonian liquid a fibrous filler even in very small concentrations causes the appearance of strong non-Newtonian effects. For this, however, a sufficiently high flexibility of fibers is required. Thus, for the ratio of the length to the diameter, l/d, of glass fibers not over 30, introducing them into a Newtonian liquid did not cause non-Newtonian effects [31]. The decisive role of l/d of the fibers in a manifestation of non-Newtonian properties of dispersion with Newtonian dispersion medium was confirmed by direct experiments [32], in which the investigators also observed a very strong divergence between flow curves at low shear rates and at sufficiently high values of l/d of dispersed fibers, as schematically shown in Fig. 11, and their convergence upon the transition to the region of high rates (or stresses) of shear.

3.5 Adsorption Effect

In conclusion to this section let us discuss the problem of interaction between a dispersion medium and filler. It was noted above that by varying intermoecular interaction on the filler's surface the yield stress may be varied over a wide range as a measure of strength of the structure being formed.

Modification of filler's surface by active media leads to the same strong variation in viscosity. We can point out as an example the results of work [8], in which the values of the viscosity of dispersions of $CaCO_3$ in polystyrene melt were compared. For $\varphi = 0.3$ and the diameter of particles equal to 0.07 nm a treatment of the filler's surface by stearic acid caused a decrease in viscosity in the region of low shear rates as compared to the viscosity of nontreated particles more than by ten times. This very strong result, however, should not possibly be understood only from the point of view of viscometric measurements. The point is that, as stated above, a treatment of the filler particles affects its ability to netformation. Therefore for one and the same conditions of measuring viscosity, the dispersions being compared are not in equivalent positions with respect to yield stress. Thus, their viscosities become different.

Nevertheless, from a practical viewpoint physical reasons for viscosity reduction during the surface treatment of the filler play a minor role; first of all the effect of viscosity reduction itself is significant.

The treatment of the filler's surface, changing its interaction with a dispersion medium, not at all obligatory causes the viscosity reduction. As it was noted during the discussion of the role of a filler's interaction with a medium the effect may turn out to be different concerning yield stress [10], and viscosity. Thus, in a number of papers [33, 34] the authors described the effect of a considerable increase of viscosity during a treatment of the filler's surface (glass fibers) by a binding agent, increasing the interaction between a polymer and filler. On the whole this effect may be treated as the formation of a stable adsorption layer, which is equivalent to an increase in the efficient volume of the filler. Assume that it took place indeed, i.e. an adsorption layer of the thickness δ was formed on the surface of the filler particles (considered for simplicity as having spherical form). Then the apparent volume of solid particles and volume content of the filler increase. Experimentally this will be expressed (for diluted suspensions of spherical particles) in the increase in the coefficient b_1 as compared with Einstein coefficient.

Let the observed value of the coefficient in linear dependence of η_{sp} vs φ be b_1'. Then we can write a simple relationship:

$$2.5\varphi_e = b_1'\varphi \tag{14}$$

where φ is the true and φ_e the apparent volume of the filler. Hence it follows that if the diameter of filler particles is d and $\delta \ll d$, then δ is expressed in the following way:

$$\delta = \frac{b_1' - 2.5}{15} d \tag{15}$$

This formula may be useful as a rheological method for determining the thickness of adsorption layer, which is formed as a result of interaction between a dispersion medium and filler, by the results of measuring the η versus φ dependence. Especially curious phenomena, connected with surface effects, arise when a mixture of fillers of different nature is used: according to concentration of an active filler the introduction of the second (inert) filler can either increase or decrease the viscosity of a multicomponent system [35].

3.6 Extensional Flow

Interpretation of data obtained under the conditions of uniaxial extension of filled polymers presents a severe methodical problem. Calculation of viscosity of viscoelastic media during extension in general is related to certain problems caused by the necessity to separate the total deformation into elastic and plastic components [1]. The difficulties increase upon a transition to filled polymers which have a yield stress. The problem on the role and value of a yield stress, measured at uniaxial extension, was discussed above. Here we briefly regard the data concerning longitudinal viscosity.

Very few experimental data, concerning measuring longitudinal viscosity of suspensions, are known. Metzner [21] carried out their critical consideration and showed that strong effects can be expected only in suspensions containing a filler in the form of long fibers. Appearance of orientation and straightening of fibers during longitudinal flow leads to a very sharp growth of resistance to deformation even when concentration of fibers is only 0.1–1%. Here a relative length (1/d) of fiber plays a great role, because longitudinal viscosity of equiconcentrated solutions increases in proportion to $(1/d)^2$.

Though experimental data on suspensions of fibers in Newtonian dispersion media give more or less regular picture, a transition to non-Newtonian viscoelastic liquids, as Metzner noted [21], makes the whole picture far or less clear. Probably, the possibility to make somewhat general conclusions on a longitudinal flow of suspensions in polymer melts requires first of all establishing clear rules of behavior of pure melts at uniaxial extension; this problem by itself has no solution as yet.

4 Elasticity

4.1 Normal Stresses

It seems quite obvious that introducing a solid filler into a polymer melt in the general case always leads to a growth of rigidity of a material. This is really so on the qualitative level. However, it is not at all obvious how this is reflected on the ability to high reversible (large elastic) deformations.

There is a widespread point of view that introducing a filler we suppress rubbery elasticity. The main proof of this point is a reduction of swelling of a stream, pressed out from a capillary tube [19, 34, 36]. Here it is a usual thought that swelling ratio is connected with normal stresses and, correspondingly, with stored elastic energy. Obviously, reduced values of swelling ratio are indeed typical for filled polymers. As regards normal stresses, the situation is more complicated.

Unfortunately, a few papers are known where normal stresses during shear flow of filled polymers were measured directly. Here an additional problem is connected with the solution of the problem what is considered a one-valued measure of elasticity of a material and under what conditions to compare the measured values of normal stresses. Moreover, the data at hand often represent rather a contradictory picture.

Thus, in papers [19, 37, 38], as in many other cases, a growth of normal stresses was noted (presented by the value of the first difference) for filled polymers as compared to base polymers, and moreover, this effect was expressed stronger when anisodiametric particles (fibers) were used as a filler. It was also noted that in all the cases the growth of normal stresses during filling was especially great in the region of low shear rates. It is important, however, to compare the rate of growth of normal and shear stresses, since shear stresses also grow with filling and also very greatly in the region of low shear rates (near yield stress).

A representative measure of rubbery elasticity of a material may be two quantities: dimensionless ratio (σ/τ) and characteristic relaxation time $\theta = \sigma/2\tau\dot{\gamma}$. According to the data of works [37, 38] when fibers are introduced into a melt, σ/τ increases (i.e. normal stresses grow faster than stresses) and θ also increases on a large scale, by 10^2–10^3 times. However, discussing in this relation the papers published earlier, we noted in the paper cited that the data were published according to which if fibers were used as a filler (as in work [37]), θ indeed increased [39], but if a filler represented disperse particles of the type TiO_2 or $CaCO_3$, the value of θ decreased [40].

Indeed, the data that with an increase in the content of spherical (or close to them) particles of the filler, the ratio σ/τ is reduced, were given in the later work [8], and here this effect was expressed stronger for the particles with smaller diameter.

Considering the problem on normal stresses at shear flow of suspensions, we should return to a specific role of anisodiametricity of filler particles. As mentioned above, nonsphericity of the filler results in the appearance of new relaxation mechanism with characteristic time D_r^{-1} and elastic effects determined by this mechanism. This theoretical conclusion was confirmed by direct experiment in paper [32], in which an appearance of normal stresses in the suspension of long fibers in Newtonian (low-molecular) liquid was observed. These normal stresses grow with an increase in concentration and flexibility of fibers just as it takes place when macromolecules are introduced in a low-molecular solvent. Thus, introducing a fibrous filler into a Newtonian liquid leads to an appearance of elasticity of the medium. This is confirmed not only by direct measurements (appearance of normal stresses), but also by all hydrodynamic phenomena like a vortex or instability for critical value of Weissenberg Number, characteristic for viscoelastic media [41]. Therefore, obviously, in the general case (and specifically for liquids with anisodiametric filler) we must distinguish between the elasticity of the medium expressed by the appearance of normal stresses at shear flow and rubbery behavior as the ability to large recoverable deformations, though the latter also leads to normal stresses.

As it is clear from the discussion, when polymers are filled the swelling of extrudate decreases, and normal stresses, measured directly, nevertheless grow. To explain the apparent contradiction Utracki [19] paid attention to the role of yield stress. Indeed, attaining, in the course of relaxation during elastic recovery (swelling), stresses of the order of Y causes further swelling to stop, i.e. during elastic recover the energy stored in a filled melt is not realized completely.

4.2 Dynamic Properties

The peculiarities of dynamic properties of filled polymers were described above in connection with the discussion of the method of determining a yield stress according to frequency dependence of elastic modulus (Fig. 5). Measurements of dynamic properties of highly filled polymer melts hardly have a great independent importance at present, first of all due to a strong amplitude dependence of the modulus, which was observed by everybody who carried out such measurements [3, 5].

A typical behavior of amplitude dependence of the components of dynamic modulus is shown in Fig. 14. Obviously, even for very small amplitudes A it is difficult to speak firmly about a limiting (for $A \to 0$) value of G', the more so that the behavior of the $G'(A)$ dependence and, respectively, extrapolation method to $A = 0$ are unknown. Moreover, in a nonlinear region (i.e. when a dynamic modulus depends on deformation amplitude) the concept itself on a dynamic modulus becomes in general not very clear and definite.

The existence of the $G'(A)$ dependence even in the region of very small amplitudes is explained by a brittle pattern of fracture of a filler's structure, so that measuring virtually frequency (and amplitude) dependences of a dynamic modulus, a researcher always deals with a material in which the structure is partially fractured.

Of course, measurements of amplitude dependence of the elastic modulus of a filled composition may, by itself, give an interesting material on thixotropic properties of such materials. Indeed, very strong effects are observed here. Thus, in a melt containing

50 weight percent of steel balls a dynamic viscosity may change up to 100 times according to the amplitude of deformation [5]. Unfortunately, at present there are no theoretical approaches, which would be able to substantiate such method. In addition the possibility to use the results of dynamic measurements in experiments was not tested in practice for the melts of linear polymers, though similar ideas were used for cured elastomers.

Though due to the fact that it is difficult to interprete amplitude dependence of the elastic modulus and to unreliable extrapolation to "zero" amplitude, the treatment of the data of dynamic measurements requires a special caution, nevertheless simplicity of dynamic measurements calls attention. Therefore it is important to find an adequate interpretation of the obtained results. Even if we think that we have managed to measure correctly the dependences $G'(\omega)$ and $G''(\omega)$, as we have spoken above, the treatment of a peculiar behavior of the $G'(\omega)$ dependence in the region of low frequencies (Fig. 5) as a yield stress is debatable. But since such an unusual behavior of dynamic functions is observed, a molecular mechanism corresponding to it must be established.

The discussion of the results of measuring the dynamic properties of filled polymers is based very often on the idea of correlation of the G'' and τ functions, which is not always expressed directly. However, due to a very sharp dependence of a dynamic modulus on the amplitude, it is not clear how to understand this correlation.

Moreover, if for pure polymer melts the correlation of the behavior of the functions $\eta'(\omega)$ and $\eta(\dot{\gamma})$ under the condition of comparing $\omega \approx \dot{\gamma}$ takes place, as a general rule, but for filled polymers such correlation vanishes. Therefore the results of measuring frequency dependences of a dynamic modulus or dynamic viscosity should not be compared with the behavior of the material during steady flow.

A possible approach to interpretation of a low-frequency region of the $G'(\omega)$ dependence of filled polymers is to compare it with a specific relaxation mechanism, which appears due to the presence of a filler in the melt. We have already spoken about two possible mechanisms — the first, associated with adsorption phenomena on a filler's surface and the second, determined by the possibility of rotational diffusion of anisodiametrical particles with characteristic time D_r^{-1}. But even if these effects are not taken into account, the presence of a filler can be related with the appearance of a new characteristic time, λ_f, common for any systems. It is expressed in the following way

$$\lambda_f = \eta_0 R^3/kT \tag{16}$$

where η_0 is the viscosity of dispersion medium, k is the Boltzmann constant, R is a typical geometrical length size of the particles of the filler (for spherical particles this is their radius). The quantity λ_f describes the velocity of filler particles as compared to the velocity of Brownian motion. Then dimensionless parameter $(\dot{\gamma}\lambda_f)$ for a steady flow or $(\omega\lambda_f)$ for dynamic loading will characterize the role of relaxation processes, associated with this mechanism, in the appearance of that or other properties of a material.

Thus, we may assume that dimensionless parameter $(\dot{\gamma}\lambda_f)$ is responsible for the appearance of non-Newtonian flow, the value of normal stresses (which appear here only due to the presence of filler particles of any form irrespective of elasticity of a dispersion medium) and the parameter ($\omega\lambda_f$) determines new, as compared to a matrix, relaxation phenomena, which appear during measurements of dynamic properties of the filled material.

As regards a qualitative pattern of influence of the filler on dynamic properties of melts of filled polymers, the situation in many respects is the similar described above for yield stress and viscosity. Indeed, the interpretation of the field of yield stress, estimated by a dynamic modulus, was given in an appropriate section.

Describing the effect of the filler and the pattern of interaction with dispersion medium (varied by a surface treatment of the filler) on dynamic modulus G' and viscosity η', we would repeat many things that were spoken about viscosity properties of the melts of filled polymers, namely speak about the growth of G' and η' with an increase in the degree of filling [42], about the reduction of G' and η' during the surface treatment of the filler, if it reduces the degree of its sintering [15] or about their growth if because of finishing the interaction between dispersion medium and filler [10], and so on.

For the purposes of constant consideration the most significant is the circumstance that all these data on the whole give the information which is equivalent or close to that obtained during measurements of rheological properties under the conditions of shear flow. Therefore a method of investigation here is determined by the taste of the experimenter and measuring technique available.

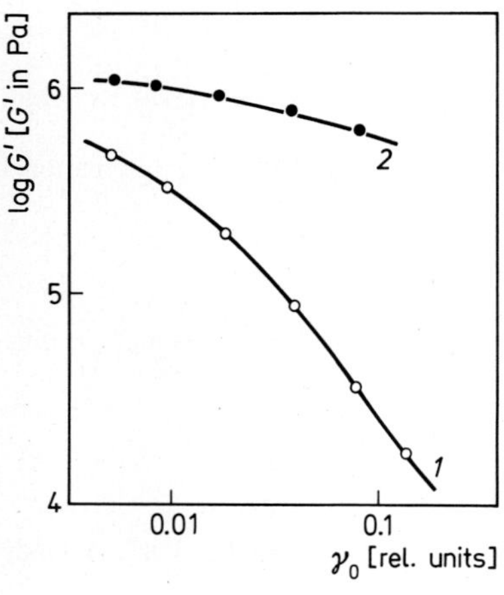

Fig. 14. Amplitude dependences (γ_0 is the amplitude of cyclic deformations) of the elastic modulus for frequency $\omega = 63\ s^{-1}$ 13% dispersion of acetylene carbon black in low- (*1*) and high-molecular (*2*) poly(isobutylene)s

5 Conclusion

A vast number of papers carried out in the field of investigation into rheological properties of filled polymers enabled us to compile a highly general picture of their manifestation. The degree of generalization is so high and reliable, that today reading a new paper on rheology of filled polymers, one often thinks that this was already mentioned, the same or very similar curve was described there and there, and analogous conclusion was made some years ago. Sometimes an impression is formed that if the names of the objects of investigation are changed, all the rest has been already described. But then it is natural to put a question: what else can be expected besides applying its general results to specific materials?

It seems that indeed the answers to many fundamental questions are obtained, at least in qualitative form. Perhaps, the most important exception are thixotropic phenomena. There are many of them and the necessary systematization and mathematical generalization are absent here. Thus, it is not clear how to describe the effect of an amplitude on nonlinear dynamic properties. It is not clear what is the depth and kinetics of the processes of "fracture-reduction" of structure, formed by a filler during deformation. Further, there is no strict description of wall effects and a friction law for a wall slip is unknown in particular.

The answers to the remaining questions are seemingly more or less known. But they are obtained in a general form as guidelines. At the same time a great number of various filled systems exists, is suggested daily and will be developed in the future. For all of them within the framework of general systematization it is necessary to determine particular indices of rheological properties, to establish their interrelation with the composition of the system and to suggest the ways for controlling rheological properties as a basis for controlling technological processes of processing these materials into articles. This is just the way to solve further problems of investigating rheological properties of filled polymer melts.

6 References

1. Vinogradov GV, Malin AYa (1980) Rheology of Polymers, Springer Verlag
2. Malkin AYa, Sabsai OYu, Verebskaya EA, Zolotarev VA, Vinogradov GV (1976) Kolloidn. Zh. 38: 181
3. Vinogradov GV, Malkin AYa, Plotnikova EP, Sabsai OYu, Nikolaeva NE (1972), International J. Polymeric Mater., 2: 1
4. Umeya K, Otaibo Y (1981) Polym. Engng. Sci. 21: 619
5. Bigg DM (1983) Polym. Engng. Sci. 23: 206
6. Yanovskii YuG, Vinogradov GV, Barancheeva VP (1986), Vysokomol. soedin., A, 28: 983
7. Vinogradov GV, Plotnikova EP, Zabugina MP, Borisenkova EK (1987) Vysokomol. soedin., B, 29: 211
8. Suetsugu Y, White JL (1983) J. Appl. Polym. Sci., 28: 481
9. Bigg DM (1982) Polym. Engng. Sci. 22: 52
10. Bretas RES, Powell RL (1985) Rheol. Acta, 24: 69
11. Stemhuis S, Loppe JPS (1982) Rheol. Acta, 21: 103
12. Platanov BE, Vdovenko NV, Ovcharenko FD, Polishchuk TA, Ermakova TS (1983) Doklady AN SSSR, 270: 136
13. Kogan EG, Pokrovskaya NV, Dolotova NN, Akopova VI, Kulichikhin SG (1988) Vysokomol. soedin., A, 30: 100
14. White JL, Czarneki L, Tanaka H (1980) Rubber Chem. Techn., 53: 823
15. Bigg DM (1983) Polymer Engng. Sci., 23: 206
16. Lakdavale K, Salovey R (1985) Polymer Engng. Sci., 25: 797
17. White JL, Tanaka H (1981) J. Appl. Polym. Sci. 26: 579
18. Tanaka H, White JL (1980) Polymer Engng. Sci., 20: 949
19. Utrecki LA (1984) Rubber Chem. Techn., 57: 507
20. Thomas DG (1965) J. Colloid Sci., 20: 267
21. Metzner AB (1985) J. Rheol., 29: 739
22. Kitano T, Kataoka T, Shirota T (1981) Rheol. Acta, 20: 207
23. de Sousa JA (1984) Advances in Rheology, ed. B Mena, Garcia-Rejon A, and Rangel-Nafaile C, Univ. Nac. Autonom. de Mexico, 3: 439

24. Chung B, Cohen C (1985) Polymer Engng. Sci., 25: 1001
25. Sanou M, Chang B, Cohen C (1985) Polymer Engng. Sci., 25: 1009
26. Ma CY, White JL, Weissert FC, Isaev AI, Nakajima N, Min K (1985) Rubber Chem. Techn, 58: 815
27. Negler RP, Mennig G (19850 Polymer Engng. Sci. 25: 395
28. Czarnecki L, White JL (1980) J. Appl. Polymer Sci., 25: 1217
29. Crowson RJ, Scott AJ, Saunders DW (1981) Polymer Engng. Sci. 21: 748
30. Miles JM, Murty NK, Modlen GF (1981) Polymer Engng. Sci., 21: 1171
31. Schröder R (1986) Rheol. Acta, 25: 257
32. Goto S, Nagazone H, Kato H (1986) Rheol. Acta, 25: 119
33. Han CD, Veghts van den T, Shets P, Haw JR (1981) Polymer Engng. Sci., 21: 196
34. Chiu WY, Hauch TC (1986) J. Appl. Polym. Sci., 32: 4663
35. Faitelson LA, Yakobson EZ (1987) Mekh. kompoz. mater. 1: 112
36. Nisimura T, Katsoka T (1984) Rheol. Acta, 23: 401
37. Kitano T, Kataoka T (1981) Rheol. Acta, 20: 390
38. Kitano T, Kataoka T, Nagatsuka Y (1984) Rheol Acta, 23: 20
39. Chan Y, White JL, Cysnagi Y (1978) J. Rheology, 33: 507
40. Minagawa N, White JL (1976) J. Appl. Polym. Sci., 20: 501
41. Goto S, Nagazono H, Kato H (1986) Rheol. Acta, 25: 246
42. Kitano T, Kataoka T, Nagatsuka Y (1984) Rheol. Acta, 23: 408

Editor: N. S. Enikolopyan
Received Mai 30, 1989

Rheology of Gas-Containing Polymer Systems

O. Yu Sabsai[1], N. E. Nikolaeva[2], M. L. Fridman[3]

The paper contains a review with original theoretical and experimental data on the basis of the rheological behavior of gas-containing systems — thermoplastic melts — during extrusion.

The more important physical aspects of the flow of gas -containing systems are analyzed for a uniphase and a biphase flow. Equations are developed for a qualitative description of these systems.

The usefulness of the proposed relationships in the engineering design of moulding equipment has been experimentally verified. A comparison is made of the pressure-volumetric flow rate characteristics with the structure of the flow and of the obtained extrudates.

Recommendations are made for calculating the boundaries of optimized flow conditions of gas-containing thermoplastic melts. These recommendations should ensure the obtaining of extrudates with a uniform fine-cellular structure and having excellent operational characteristics. A good agreement has been obtained for experimental and theoretical results.

1 Introduction 100

2 Formation of the Free Gas Phase 102

3 Bubble Dynamics 106

4 Single-Phase Flows 109

5 Two-Phase Flows 113

6 Flow Structures 117

7 Conclusion 120

8 References 121

[1] 105554 Moscow, Nizhnyaa Pervomaiskaya ul. 24, apt. 218
[2] 121352 Moscow, Davydkovskaya ul. 10, bd. 5, apt. 18
[3] 103062 Moscow, ul. Chernyshevskogo 41, apt. 70

Advances in Polymer Science 96

1 Introduction

Gas-filled plastics are polymer materials — disperse systems of the solid-gas type. They are usually divided into foam plastics (which contain mostly closed pores and cells) and porous plastics (which contain mostly open communicating pores). Depending on elasticity, gas-filled plastics are conventionally classified into rigid, semi-rigid, and elastic, categories. In principle, they can be synthesized on the basis of any polymer; the most widely used materials are polystyrene, polyvinyl chloride, polyurethanes, polyethylene, polyepoxides, phenol- and carbamideformaldehyde resins, and, of course, certain organosilicon polymers.

Gas-filled plastics, which combine lightness with relatively high specific strength and fair thermo-, sound- and electroinsulation properties, are coming into increasing use in various domains of technology. The combination of such properties stems primarily from structural individualities of foam plastics, which are essentially filled polymer compositions, where air or another gas acts as a filler. The characteristics of the filler, notably density and tensile strength, are several orders lower than analogous indexes of the polymer matrix. Furthermore, fillers of this type possess considerable compressibility and the ability to dissolve in a polymer medium.

The introduction of large gas phase volumes into the polymer alters the physical characteristics of the material: volume weight, permeability to fluids and gases, and physico-mechanical properties. Moreover, the properties of the polymer matrix itself are changed (owing to orientation effects, supermolecular structure of the polymer in the walls, ribs and tension bars of cells), which drives up the value of specific strength on impact, and results in anisotropy of elasticity.

Typically, large-scale gas filling makes the main characteristics of foam plastics — coefficients of heat and temperature conductivity, dielectric permeability, and the tangent of the dielectric loss angle — totally independent of the chemical structure of the original polymer [1].

The synthesis of gas-filled plastics from high polymers or their mixtures involves technological procedures and equipment that are largely similar to those used to produce corresponding monolith plastics (pressing, extrusion, pressure molding, baking, etc.). The foaming process occurs at high temperatures, which are near or beyond the temperature of viscous flowing of corresponding polymers.

Technological breakthroughs in the production of articles made of gas-filled polymers have until recently out-distanced the development of scientific views on physico-chemical regularities governing polymer foam formation. The empirical approach is still dominant in the creation of novel materials and processes for their synthesis and processing into articles. Hence, several problems require the elaboration of concrete and general physico-chemical regularities, which determine formation and behaviour of polymer foams, with the use of assorted methods from various disciplines: physical and colloid chemistry, physico-chemical mechanics, rheology, physics and mechanics of multicomponent media, physics of surface phenomena, and phenomena of mass- and heat transfer.

The following materials serve as ingredients for foam plastics: the polymers, blowing agents, hardeners, plastifiers, lubricants, stabilizers, fillers, and other substances.

A cellular or porous structure of foam plastics is produced with gas- or foam-forming agents. The quality of the resulting material depends on choosing the right agent, whose type and nature determine the method by which foam plastics are to be made, and, of course, technological parameters.

The methods for obtaining foam plastics are divided, by convention, into chemical, physical, and mechanical methods. The essence of each process, along with the classification of blowing agents, their chemical nature and requirements governing their selection are described in some detail in the literature. Below we shall examine a process of chemical foaming, which amounts to decomposition of the blowing agent with generation of the gases that saturate the polymer melt. Foaming with chemical blowing agents (notably, organic ones) yields foam plastics with the most even structure and also the smallest cells; such plastics have optimum physico-mechanical properties.

It should be emphasized that the following analysis pertains only to conditions under which the gases contained in the polymer are far from their critical point, and are not condensed into liquids.

Rheological behaviour of melts with condensed gases is reported in [2–6].

Figure 1 explains the goal of this review [7]. The Figure shows that there exists a narrow range of pressures (P) and volumetric flow rates (Q), in which it is possible

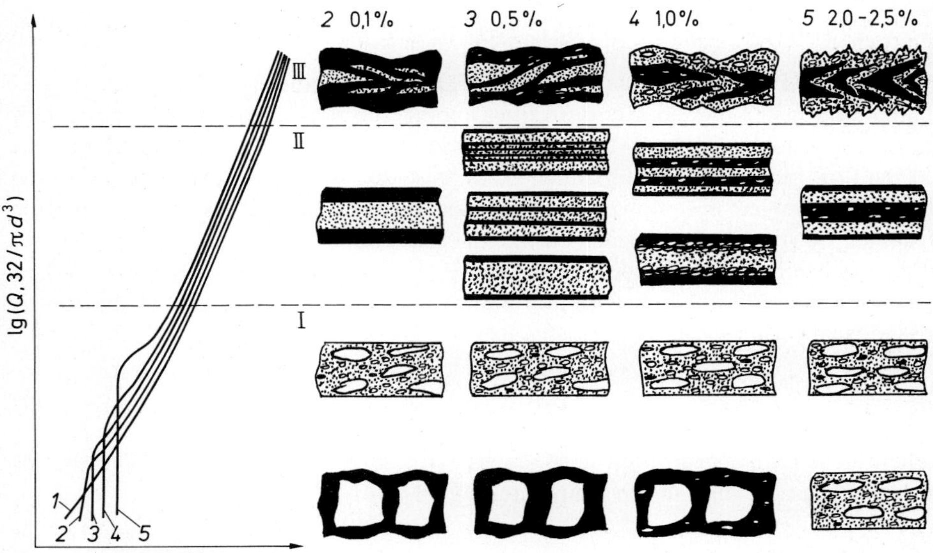

Fig. 1. Axial Sections of Extrudates of Gas-containing LDPE Melts Depending on Mass Content of Blowing Agent (azodicarbonamide, CXZ-21, from 0.1% to 2.5%) and Volumetric Flow Rate (T = 190 °C)

to obtain an undistorted small-cellular extrudate of foam. Here, it is of tremendous importance for the engineer to establish the limits of this range in advance, to predict the P — Q characteristic of the thermoplastic material which contains gas at specified parameters and say what equipment is necessary for the process. It is to the analysis of this problem that the review is devoted. For this purpose, kinetics of blowing agent decomposition will have to be examined, along with the PVT diagram of gas-containing polymer melts, and the individualities of rheological behaviour of thermoplastics which contain gas. All these problems are, in our opinion, interesting in themselves, particularly to specialists in polymer physics and chemistry.

2 Formation of the Free Gas Phase

As a rule, the blowing agent does not undergo pre-decomposition when foam articles are produced under actual technological conditions. Polymer movement in technological equipment takes place simultaneously with the decomposition of the blowing agent, introduced into the thermoplastic. This alters both overall gas content and the ratio between free gas and gas dissolved in the polymer. Furthermore, temperature and pressure along the melt flow axis are also changed. Gases are known to dissolve in polymer melts. Two types of flows are possible in one and the same technological process (even in the same molding machine — along the melt flow axis), depending on the value of hydrostatic pressure at a given point [8]:

a) gas is totally dissolved in the melt, and the flow is single-phase;

b) a two-phase flow, when part of the gas escapes from the solution in the form of bubbles.

The value P_{cr}, critical pressure, is a fundamental parameter of the nature of gas-containing melt flows; below this critical point, the flow becomes two-phase. This point was established by C. D. Han and C. A. Villamizer [8] by direct observations of the flow in a transparent channel. Data presented by these authors pertain to the case of incomplete decomposition of the blowing agent prior to its entrance into the channel. Obviously, P_{cr} depends on the amount of gas.

Let us examine the connection between P_{cr} and gas solubility parameters. When gas content in the polymer does not change with time, gas solubility in the melt is described by Henry's Law:

$$C_g = H_T \cdot P_g \tag{1}$$

where C_g is gas concentration in the melt; H_T is solubility; P_g is partial pressure of dissolved gas. H_T depends on temperature [9–11]:

$$H_T = Ae^{-K/T} \tag{2}$$

In Eq. (2) T is the absolute temperature, whereas A and K are constants for the specific polymer-gas couple. According to [11–15], their values for the couple PE + N_2 are: A = 0.313464; K = —478.589; PE + CO_2: A = 0.114725; K = 403.0227; PS + N_2: A = A = 0.007645; K = 856.423 (A is cited in cm^3/cm^3 bar; K is in Kelvin).

Part of the gas can escape from the solution at a specific concentration and a fixed temperature, as the pressure level falls to under $P < P_g$. This takes place in two phases: appearance of nuclei, and growth of bubbles of the free gas phase. Thermodynamic conditions for stable nucleation are formulated in [1]. They are analogous to the conditions for starting the boiling of low-molecular liquids. The following changes take place in the isobar-isothermic potential of the system ($-\Delta Z$) as a single bubble nucleus is formed:

$$-\Delta Z = \sigma \cdot S - \Delta\mu \cdot i$$

where the product of surface tension by bubble surface area $\sigma \cdot S$ is a change in surface energy, and the product of the change in chemical potential in the system by the number of gramme-moles of gas in the bubble $\Delta\mu \cdot i$ is a change of chemical potential in the bubble nucleus.

For an ideal gas,

$$\Delta\mu = RT \ln (P_g/P)$$

and

$$-\Delta Z = 4\pi[r^2\sigma - (r^3/3\,Vm) \cdot RT \ln (P_g/P)] \tag{3}$$

where V_m is the volume occupied by one gramme-mole of gas.

From Eq. (3), we get the critical nucleus radius

$$r_{cr} = 2\sigma V_m/RT \ln (P_g/P)$$

Usually, critical radii $r_{cr} = 10^{-2} + 10^{-1}\ \mu$ under processing conditions. The probability θ of random appearance of such a nucleus in a volume unit in a time unit, calculated from Eq (3), is close to zero:

$$\theta \sim \exp\left(-\frac{\Delta Z_{cr}}{RT}\right) = \exp\left[\frac{-16\pi\sigma^3 V_m^2}{3(RT)^3 \ln^2 (P_g/P)}\right]$$

This closeness of θ to zero explains the existence of a gas-oversaturated solution area in the polymer melt, when $P < P_g$, but the entire volume of gas remains in the solution. The degree of oversaturation, particularly upon free foaming (not in flow) can be 2- to 3-fold. In real polymer compositions, there are always solid admixtures, which have poor wetting areas. This reduces the degree of oversaturation at the interface melt-molding tool. Moreover, bubble nuclei can result from fragmentation of gas bubbles in the polymer [16]. Another factor that promotes the formation of bubble nuclei is the presence of localized "hot points" in the polymer melt; they act as nucleation centres. "Hot points" appear either after a chemical reaction in the melt polymer [17], or in overheated areas on the surface of metal equipment [18]. Density of nucleation can be improved via introduction of various agents that reduce tension of the polymer [19].

In flows of gas-containing polymer melts, local eddies can form to reduce the oversaturation degree, after the flow changes direction, or the channel — its cross-section.

All this points to the conclusion that $P_{cr} = P_g$ at least for the flow of gas-containing polymer melt.

Hence, if the chemical blowing agent (CBA) is fully decomposed, Eq. (1) gives us

$$P_{cr} = P_g = C_g/H_T \tag{4}$$

But $C_g = CG$
where C is CBA concentration, and G is yield of gas; subsequently

$$P_{cr} = CG/H_T$$

or, more accurately, taking into account temperature, polymer mass and atmospheric pressure, P_0 [20, 21]:

$$P_{cr} = [GCT_i/H_T T_0(100 - C)] + P_0 \tag{5}$$

(T_0 is standard (room) — temperature; C is weight content of CBA in per cent).

If the flow is accompanied with CBA decomposition, the G value in Eq. (5) should be substituted with its time function, G(t). In the general case, thermal decomposition of a solid substance with gas emission is a heterogeneous topochemical reaction [22]. Kinetic curves of such reactions are "S"-shaped; the curves representing reaction rate changes in time pass a maximum. At unchanging temperature, the G(t) function for any CBA is easily described with the Kolrauch exponential function [20, 23, 24]:

$$G(t) = G(1 - e^{-\alpha t^p}) \tag{6}$$

In Eq. (6), α and p are constants. The p value has a weak linear dependence on temperature; α is exponentially dependent on T [23, 24].

We have no knowledge of any work involving the study of the dependence of α and p on hydrostatic pressure.

Useful constants that can be measured in the course of the experiment are: time of full foaming, t_f and time of CBA half-decomposition, $t_{1/2}$. Both values fall exponentially with temperature.

Experimentally observed t_f and $t_{1/2}$ depend both on temperature, and on the type of CBA and the medium where it decomposes [1]. There exist numerous decomposition activators [1, 25], which change activation energy and kinetic constants of CBA decomposition. Hence, constants in kinetics Eq. (6) (or t_f; $t_{1/2}$) should be specially determined for each polymer-CBA couple, in order to calculate α and p with t_f and $t_{1/2}$ [20, 21].

Eqs. (5) and (6) give us [20, 21]:

$$P_{cr}(t) = [GC(1 - e^{-\alpha t^p})T_i/H_T(100 - c)\,T_0] + P_0 \tag{7}$$

(here, C is weight % concentration of CBA), Fig. 2 proves the theory's applicability. Kinetic constants and their temperature dependence have been determined from experimental data [21], and work [8] for concentration of CBA = 0.8 mass%. The CBA decomposition time resulted from experimentally measured P_{cr} [8] (dotted line). Solid lines represent calculation results for other foaming agent concentrations, assuming that t_f and $t_{1/2}$ do not depend on concentration.

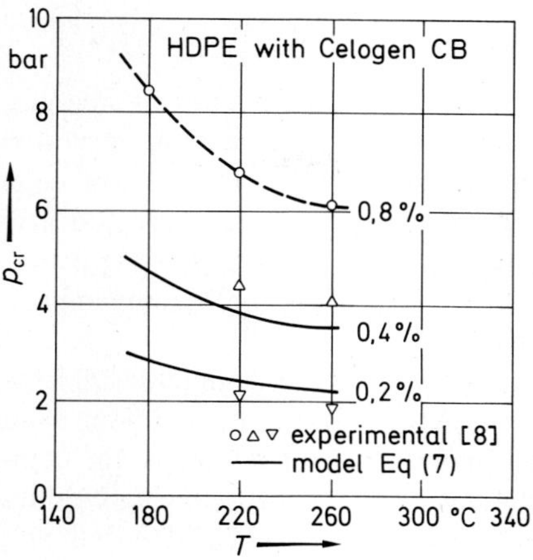

Fig. 2. Dependence of Non-Equilibristic Critical Pressure (P_{cr}) on Temperature

Let $P < P_{cr}$. Part of the gas has escaped from the solution in the form of the free phase — bubbles. On the basis of the above, we can always calculate the amounts of dissolved and free gas [20], provided that P is lower than P_{cr}, and P = constant. When CBA decomposes fully:

$$C_{g\,dis} = H_T P \tag{8}$$

$$Cg_{free} = H_T(P_{cr} - P) \tag{9}$$

P_{cr} can be calculated with Eq. (5) for Eq. (9).

Should CBA fail to decompose fully, it is easy to draw up an equation (using Eq. 7), to determine the value of equilibristic volume on foaming, i.e. the volume on attaining which bubbles quickly reach equilibrium dimensions, making kinetics of their growth negligible. This equation was derived in [21], and tested experimentally [21] for LDPE + N_2 and HDPE + N_2; also, for PS + N_2 [26]:

$$V_\infty = V_{melt} + \frac{P_0}{P_i}\left[GC\,\frac{T_i}{T_0}\,(1 - e^{-\alpha t^P}) + H_T(100 - c)\,(P_0 - P_i)\right] \tag{10}$$

In Eq. (10), V_∞ is equilibrium volume of foamed polymer, V_{melt} is volume of thermoplastic melt at experimental temperature T_i; P_i is foaming pressure. Other symbols are the same as in previous formulas.

It should be pointed out that Eq. (10) can correctly represent experimental results (cited papers recorded reasonable agreement between theory and experiment) only when diffusion of gas from bubble to bubble can be ignored. When the bubble is small, surface Laplace pressure $P_L \gg P_i$, and coalescence of bubbles occurs in such a way as to make the volume of the resultant bubble greater than the sum of the original bubble volumes [27].

3 Bubble Dynamics

Hydromechanic rheologists are naturally interested not in the amount of free gas, but in its volume, V_{free}, or, better still, the volume and size distribution of generated bubbles. These things depend on a host of factors. From the point of view of most investigators of gas-containing media, their dependence on time, pressure and prehistory of pressure changes, is of the greatest interest. This task is interesting per se, however, its applications go far beyond our present subject. This explains the impressive amount of works dedicated to this problem.

Let us start examination of bubble dynamics with a quotation from [28]: "A rigorous theoretical analysis describing the bubble dynamics during mold filling is an extremely difficult task. Difficulties arise from, among many other things, the complexity of the flow geometry in general, interactions among many growing bubbles (e.g.) the coalescence of two or more bubbles of different sizes to yield a larger bubble), the non-sphericity of growing bubbles, and the rheologically complex nature of molten polymers".

Let us start with a model situation: there is a single bubble in the liquid, the gas is insoluble, and there is no flow. Internal pressure P_{int} in the equilibrium bubble is in this case counterbalanced by external pressure P_{ext} and surface Laplace pressure $P_L = 2\sigma/r$ (σ is surface tension):

$$P_{int} = P_{ext} + P_L \tag{11}$$

Let us alter external pressure from P_{ext} to P'_{ext}. As can be seen from Eq. (11), two essentially different situations may follow:

a) P'_{ext}; $P_{ext} \ll P_L$ (fine bubbles) — in this case, bubbles radius remains unchanged, and the bubble won't "feel" the pressure change. This bubble is "hard".

b) P'_{ext}; $P_{ext} \gg P_L$ — "soft" bubbles. It should be noted that P_{int} is proportionate to the quantity of gas contained in the bubble, and inversely proportionate to its volume.

$$P_{int} = \frac{\alpha}{r^3} \tag{12}$$

where α = const., if, of course, there is no diffusion. Then, the dynamic equation, which describes the alteration of bubble radius in a Newtonian liquid takes the following form (inertialess approximation):

$$\alpha/r^3 = P_{ext} + 4\eta\frac{\dot{r}}{r} \tag{13}$$

In Eq. (13), medium resistance to bubble compression-decompression depends on viscosity η, and is described by the second member in the right-hand part of the equation. It should be mentioned at this point that bubble growth in a Newtonian liquid was originally examined by the Soviet physicist Y. I. Frenkel [29], in a rarely cited work published in 1946.

Should the medium be rheologically complex, the second member in the right-hand part of Eq (13) should be substituted with a functional φ dependent on time and rheological properties of the medium. When Laplace pressure cannot be ignored, $2\sigma/r$ should be added to the right-hand part of Eq. (13). When external pressure is altered, P'_{ext} is substituted with the $P'_{ext}(t)$ function. This gives us, in general form:

$$\alpha/r^3 = P'_{ext}(t) + \varphi(t, \dot{r}/r) + 2\sigma/r \tag{14}$$

Yet, Eq. (14) does not describe the real situation. It must also be taken into account that gas concentration differs in the solution and inside the bubble and that, consequently, bubble growth is affected by the diffusion flow that changes the quantity of gas in the bubble. The value of α in Eq. (14) is not a constant, but a complex function of time, pressure and bubble surface area. To account for diffusion, it is necessary to translate Fick's diffusion law into spherical coordinates, assign, in an analytical way, the type of function — gradient of gas concentration near the bubble surface, and solve these equations together with Eq. (14).

The series of works by C. D. Han and Hee Ju Joo [28, 30] stands out from the papers devoted to concretization and solution of equations like Eq. (14). They also supply an analysis of past works in the sphere, with references (cf. also the work by C. D. Han [31]). The most thorough programme has been carried out in [28]. The authors described rheological properties of the medium with the De Witt model. The model is a generalization of the Maxwell equation, with the usual time derivative of tension substituted with the Jaumann derivative. Unfortunately, the De Witt model fails to describe, even qualitatively, the rheological properties of polymer melts in the area of their non-linear, viscoelastic behaviour (cf., for instance, chapter 6.4 in [32]). Nevertheless, the results of [28] are very important, since the authors managed to obtain the level of figures in their analysis when they determined the effect of diffusion rate and medium elasticity on the kinetics of bubble growth; in the asymptotic phase, bubble growth is totally controlled by diffusion, whereas dr/dt is roughly proportionate to the square root of the diffusion coefficient; in the initial phase of bubble growth,

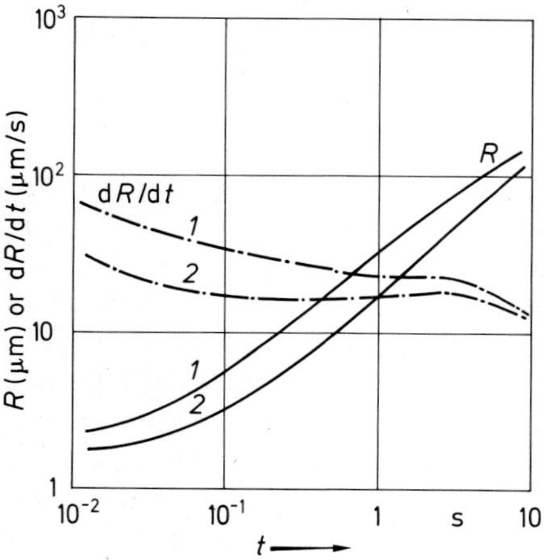

Fig. 3. Effect of Melt Viscosity on Bubble Growth During Mold Filling: (*1*) $\eta_0 = 4 \times 10^3$ N · s/m²; (*2*) $\eta_0 = 4.10^4$ N · s/m². It is assumed that $\tau_{rr,0} = 3.2 \times 10^3$ H/m², and that initial mold pressure is 1.77 atm (1.79×10^5 N/m²). Other system parameters: $\varrho = 0.88$ g/cm³; $K_p = 4.26 \times 10^{-9}$ m²/N; $D = 5.5 \times 10^{-6}$ cm²/s; $\sigma = 2.8 \times 10^{-2}$ N/m (28 dyn/cm); $Co = 2 \times 10^{-3}$ g/g; $\eta_0 = 4 \times 10^3$ N · s/m²; $\lambda = 0.90$ s

viscoelastic melt properties are important. Figure 3 illustrates the influence of viscosity (the Figure comes from [28]). Alteration of viscosity by an order alters dr/dt only 1.5 to 2 times. Close to the end of [28], the authors themselves point to the incomplete nature of their analysis: "Non-isothermic molding is common in industry. There is a need for both theoretical and experimental studies of gas bubble growth in non-isothermal conditions". We believe that the programme of research (theory) outlined in the final part of this quotation, would be difficult to implement for at least three reasons: for one thing, all starting constants and functions in Eq. (14) and the diffusion equation are subject to the influence of temperature. For another, account must be taken of the unusual mechanisms of formation, growth and movement of gas bubbles at the interface, melt — crystallized or vitrified polymer. We shall examine bubble behaviour at solid-liquid phase interface in greater detail later. Finally, calculation of non-isothermal flow (upon filling a mold, for instance) in a channel of random geometry, where temperature and pressure changes in time and space, depend on viscosity, medium heat- and temperature-conductivity (which in turn depend on the foaming degree), is a near-impossible task.

M. Amon and C. D. Denson [33–34] attempted a theoretical and experimental examination of molding a thin plate from foamed thermoplastic. In the first part of the series [33] the authors examined bubble growth, and in the second [34] — used the obtained data to describe how the thin plate could be molded with reference to the complex situation characterized in our third note. Here, we are primarily interested in the model of bubble growth per se, and, of course, the appropriate simplification proposals [33]. Besides the conditions usual for such situations: ideal gas, adherence to Henry's law, negligible mass of gas as compared to mass of liquid, absence of inertia, small Reynolds numbers, incompressibility of liquid, the authors postulated [33] several things that require discussion:

1. Formation of (nuclear) bubbles proceeds in a heterogeneous fashion. This matter has already been discussed — in part — in Sect. 2. Direct observations of gas-containing thermoplastic flows [8] have shown that this theory is quite sound and is, in fact, confirmed experimentally. In Ref. [8], the area diving the one-phase and two-phase type of flow along the axis is small. More proof to the heterogeneous nature of bubble formation is cited in [35, 36]. These works studied destruction of uncured monodisperse rubbers under triaxial extension. A thin rubber film was stretched between 2 holding punches, one of which was transparent. The stressed state of the sample attained in these experiments — triaxial extension — corresponds to bubble formation conditions, when the bubble results from the nucleus on formation of the free gas phase on changing pressure in gas-containing melt (pressure changes from $P > P_{cr}$ to $P < P_{cr}$). During fracture, one-two bubbles appeared at first, followed by, after a certain time (induction period), immediate formation of a multitude of bubbles in the entire area visible through the transparent puncher.

2. Bubbles are formed instantaneously. This conclusion made in [33] is based on estimates taken from earlier works [37]. As seen from the above cited works by S. E. Sosin et al., this is not always true: viscoelastic liquids under triaxial stretching stress are not destroyed instantly. The existence of an induction period may produce a considerable effect on foam growth kinetics upon free foaming, when pressure is lowered instantaneously from $P > P_{cr}$ to $P < P_{cr}$ in a melt with dissolved gas. However, it would appear that microfaults in polymer melts, which are caused by factors

described here in Sect. 2, cause instant bubble formation as soon as pressure is lowered during the flow of gas-containing melts, reaching a level less than P_{cr}.

These two considerations allowed M. Amon and C. D. Denson to avoid difficulties pertaining to the assignment of concentration gradients near the bubble wall. The authors called their model "the cellular model". Setting the quantity of bubbles, they placed each bubble in correspondence with a spherical cell of surrounding liquid with a mass equal to the ratio of the entire liquid mass to the overall quantity of bubbles. This made it possible for them to solve the problem of bubble growth in this cell.

3. The authors of [33] considered their liquid Newtonian. Going back to Fig. 3 and its discussion, it becomes clear that this theory is fully applicable in qualitative and even semiquantitative analysis of thermoplastic foaming.

4. The properties of molten thermoplastic are independent of the concentration of gas dissolved therein. This assertion is probably true, and very original. It clearly requires special discussion.

4 Single-Phase Flows

L. L. Blyler and T. K. Kwei [39] proposed the direct opposite (to 4). In their reasoning, they proceeded from the known and generally acceptable Doolittle equation, which puts liquid viscosity in exponential dependence on the inverse value of the free volume of the latter. According to [39], gas has a volume of its own, the value of which it contributes to the free volume of the polymer when it dissolves therein; as a result, viscosity falls. The theoretical formula obtained by the authors was experimentally confirmed in the same work. The authors measured pressure values at the entrance of cylindrical capillaries, through which melts of both pure polyethylene, and polyethylene with gas dissolved in it, extruded at a constant rate.

Before subjecting L. L. Blyler and T. K. Kwei's work to criticism, let us point out its strong points. First and foremost, this concerns the question: How does gas behave after dissolving in the melt? Analysis of gas solutions in low-molecular liquids is, evidently, based on the same grounds as the one for solutions of low-molecular liquid vapours with sufficiently large molecules in polymer melts.

The authors of [40] used L. L. Blyler's and T. K. Kwei's formula to process experimental data [41, 8] and obtained good correlation between theory and experiment. In all the processed experiments viscosity was established in accordance with pressure at channel input. To describe data presented by C. J. Ma and C. D. Han [2–5], who studied freon-containing polymer melts, the same paper supposed that the entire volume of gas is expended in part on the increase of the free volume of the composition, and that the occupied volume also changes in its presence. This consideration made it possible for the authors of [40] to attain fair correlation between theory and practice. This makes, in our opinion, the ideas expressed in [39, 40] worthy of the most serious attention, however critical the following evaluation of these works may appear to the reader.

When they calculated viscosity from pressure at the die entrance, the authors thought pressure along the channel axis to change in a linear way. This is totally incorrect when the pressure reaches the P_{cr} point at some point inside, and along the axis of the channel. Furthermore, the value of the pressure at the die exit cannot be ignored [42].

Consequently, in order to study rheological behaviour of gas-containing polymer melts, and to determine viscosity accurately, it is necessary to measure pressure along the channel axis and to determine an apparent viscosity the way it is usually done in classical hydromechanics — with local pressure deviations along the channel dP/dl. This plan was followed in [20, 21] with LDPE + N_2 and HDPE + N_2 and in [43] with PS + N_2. The experiments were carried out with a capillary rheometer "Rheograph 2000" (from "Gettfert", West Germany) at different plunger-movement speeds.

The flow of gas-containing melts was studied as they passed through a 200 × 10 × 2.0 mm rectangular slit die. Four pressure transducers (Dynisco) were installed along the flow axis in the capillary, whereas the 5th transducer was placed in the viscometric reservoir near the entrance to the capillary. In all cases, the kinetics of CBA decomposition had been studied in advance; prior to activating the plunger, materials were kept in the viscometric reservoir for a longer time than was required for total decomposition of the blowing agent at a present experimental temperature. Figure 4 cites the usual experimental results. Pressure transducers are marked by arrows on the horizontal axis. The dotted line marks P_{cr}, which was calculated with Eq. (5). Reduced volumetric flow rate is marked $\dot{\gamma}'$ ($\dot{\gamma}' = 6Q/bh^2$). Figure 4 shows that in the single-phase flow area at $P > P_{cr}$, the pressure profile is linear and parallel to the pure melt profile. In this way, introduction or addition of dissolved gas does not alter and, consequently, leaves the apparent viscosity unaffected.

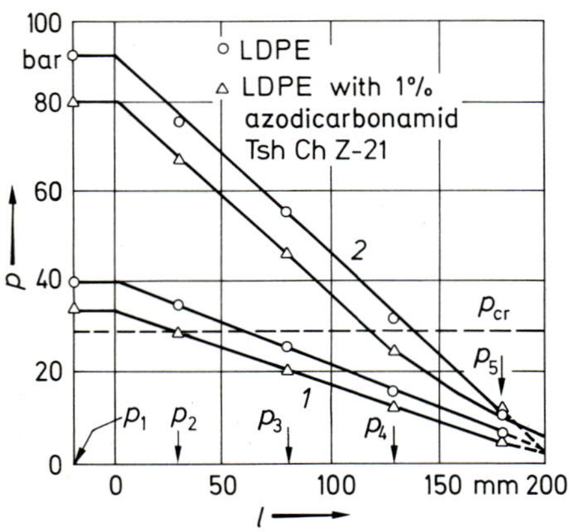

Fig. 4. Axial Pressure Profiles in Rectangular Slit Capillary (l = 200 mm, h = 2 mm, b = 10 mm), T = 190 °C; γ': 3.39 s^{-1} (*1*); 19.96 S^{-1} (*2*). Time of exposure period to extruding — 1200 s

L. L. Blyler and T. K. Kwei processed their own experimental data in an incorrect way: polyethylene melts with dissolved nitrogen have the same viscosity as pure melts. Dissolved nitrogen cannot change the value of the free volume of the PE melt; hence, L. L. Blyler's and T. K. Kwei's formula is incorrect.

Figure 4 also shows that pressure does not change along the channel axis in the $P < P_{cr}$ area in a linear pattern. Pressure at the die exit is much greater than atmospheric pressure; it would therefore be a mistake to calculate viscosity the way it was done by L. L. Blyler and T. K. Kwei and their followers [40].

Summing up, we can state that work [39] is a combination of erroneous theory and incorrect processing of experimental data, which yielded a value for mean dissipative function, which the authors mistook for viscosity. Yet, this function can be used to calculate pressure required for extrusion of gas-containing polymer melt through the capillary.

Before [39] came out, the authors of [44] claimed that dissolved gas molecules were so small that they could not alter the value of the free volume. Regardless of interpretation of the term "free volume" [45], concepts of melt structure [46] and size distribution of free holes [47], it has been estimated that an average hole contains thousands or tens of thousands of gas molecules. That is, the gas "pumped" into the polymer behaves as free gas from the point of view of statistical thermodynamics. It is also noteworthy that the life span of the hole [48, 49] is much greater than characteristic gas relaxation time; hence, melt mobility cannot effect the thermodynamic state of the gas it contains, i.e. the melt is a rigid porous medium as far as gas is concerned. Let us go back to Fig. 4. The pressure level in the area of single-phase melt flow is lower than that in a pure melt by a value ΔP. This value is governed by gas concentration, and is almost independent of the reduced volumetric flow rate, $\dot{\gamma}'$ [21]. In view of the above, the authors of [21] described ΔP as partial pressure of the gas dissolved in the melt and occupying the free volume of the polymer. The energy of the gas "pumped" into the polymer acts as a source of an additional force that push melt through the channel. This energy is freed when gas passes to free state from the solution. For an ideal gas, N. E. Nikolaeva, O. Yu. Sabsai and M. L. Fridman [20, 21], came up with the following:

$$\Delta P = \left(GC \frac{T_i}{T_0}\right)\left(\frac{\varrho_{melt}}{100 - c}\right)\left(\frac{P_0}{f_{T_i}}\right) \tag{15}$$

In Eq. (15), all symbols are the same as in preceding formulas, and f_{Ti} is the share of melt free volume at experimental temperature. Its value is easily determined from the known approximated formula (cf., for instance, [50]).

$$f_{Ti} = f_g + \alpha(T_i - T_g) \tag{16}$$

In Eq. (16), f_g is the share of free volume at glass-transition temperature T_g, universal constant equals 0.025 approximately; α is the difference between coefficients of volumetric expansion of the polymer before and after the glass-transition point. Equation (15) was tested in [21] on flows of gas-containing melts of polyethylenes of high and low density in round (l/d: 5/0.05; 10/1; 30/1; 40/4), and rectangular slit (l/h; 200/2; 85/0.5) channels. Table 1 sums up the results of the theory-experiment correlation. Azodicarbonamide was used as the foaming agent. Considering that Eq. (16) is approximated, and that experimental errors are always possible, the accuracy of the ΔP prediction with Eq. (15) is indeed impressive.

This means, of course, that an energy equation is necessary for the description of gas-liquid flows, along with the usual equations of movement and continuity. Transformation of the internal energy of dissolved gas into medium movement energy is what causes the observed pressure drop at the die entrance, e.g. the apparent decline in the amount of energy required to transport the gas-containing melt.

However, we believe that Eqs. (15) and (16) can prove quantitatively correct only for experimental data at melt temperatures that are far from the glass-transition point,

Table 1. Comparison of Eq. (15) to experiment [20]

No.: Polymers T °C	Blowing agent concentration, % mass	Number of experiments on various channels and various volum. flow rates	P_{cr} (bar) in accordance with Eq. (5)	ΔP_{exp} (mean) bar	ΔP_{calcd} Eq (15), bar	$\% = \frac{\Delta P_{calc} - \Delta P_{exp}}{\Delta P_{calc}} \cdot 100$
1. LDPE 10802-20 (USSR) 190°	0.1	7	3.98	1.25	1.22	2.7
2. LDPE 10802-20 (USSR) 190°	1.0	8	31.14	10.09	12.32	18
3. LDPE 10802-20 (USSR) 190°	2.0	4	61.92	21.84	24.89	12.3
4. HDPE Lupolen BASF, 230°	0.166	11	4.98	5.26	5.41	3
5. HDPE Lupolen BASF, 230°	0.3	7	9.97	10.78	9.79	10
6. HDPE Lupolen BASF, 230°	0.6	10	18.0	19.29	19.64	2
7. HDPE Lupolen BASF, 230°	1.0	8	31.15	32.92	32.85	0.2

and for simple gases. It appears that near the glass-transition point, as well as for large gas molecules and/or experimental conditions which bring gases near the critical point, account must be taken of changes in proper and free melt volumes as gases are dissolved therein, in accordance with L. L. Blyler and T. K. Kwei [39] and Yu. M. Kukharsky et al. [40].

In conclusion, it should be pointed out that recently [51], a considerable growth of specific fluid volumetric flow rates was discovered near the saturation pressure on filtration of the solution of CO_2 in normal heptane and gas-liquid fossil carbohydrates (oils). A possible explanation of this effect can be found in the above theoretical discussion. Finally, going back to M. Amon and C. D. Denson's work [33], which was discussed at the end of Sect. 4, let us admit that their thesis No. 4 (melt properties as regards thermoplastic itself do not depend on gas concentration) is quite correct and in good correlation with experimental results [21].

5 Two-Phase Flows

Equations (5) and (15) yield another important equation: a formula for calculating point coordinates along the channel, where two-phase flow originates — l_{cr} [20]:

$$l_{cr} = l\left(1 - \frac{\Delta P}{P_{ent_0}} - \frac{P_{cr}}{P_{ent_0}}\right) \tag{17}$$

where l is channel length, P_{ent_0} is the pressure at channel entrance on flow of pure melt. P_{cr} in Eq. (17) can also be determined on partial decomposition of the blowing agent in accordance with Eq (7); then, $l_{cr} = l_{cr}(t)$.

In an early work [52], A. M. Kraynik proposed another formula for determining l_{cr}. According to [52], densities and viscosities of foamed and unfoamed melt have to be known in order to calculate l_{cr}. Determination of these parameters is by no means easy, which is why Eq. (17) appears better than the one proposed by A. M. Kraynik.

When P_{cr} and l_{cr} are attained, a part of the gas evolves from the melt, and the flow becomes two-phase. In their analysis of gas-liquid flows, mathematicians describe this mixture as a multiphase system [53]. Equations for the movement of each phase are set down separately, along with continuity equations for each phase and borderline conditions at phase interfaces. The resultant complex system of equations has a solution, although in many cases it is difficult to prove the theorem of the solution's existence. However, many problems pertaining to flows of low-molecular liquids containing gas, even when they are accompanied by mass-exchange among phases, can be posed and solved. The problem becomes much more difficult on transition to gas-containing polymer liquids, owing to the necessity of accounting for non-linear viscous, and frequently — nonlinear viscoelastic, rheological behaviour of polymer melts.

Here is a brief account of the main experimental and theoretical results obtained in the analysis of the rheological properties of low-molecular (Newtonian) liquids with gas bubbles:

1. Viscosity, η, on small volume concentration of bubbles, φ, and a large distance between bubbles, is calculated as follows:

$$\eta = \eta_0^*(1 + \varphi) \tag{18}$$

where η_0^* is pure liquid viscosity.

This formula was first obtained by G. I. Taylor. A more generalized situation is examined in [54].

2. Also worth mentioning is the discussion between G. I. Taylor and R. Davies [55], which results in the following:

2a) Introduction of gas bubbles into non-compressible Newtonian liquid gives the latter a new property — volumetric viscosity $\varkappa$:

$$\varkappa = 4/3\eta_0^* \frac{1}{\varphi}$$

2b) Due regard to elastic compressibility yields volumetric viscoelastic properties of gas-containing liquids.

2c) The dependence of volumetric viscosity on gas concentration has a maximum. For water, $\varkappa_{max} = 6.7 \times 10^3 \times \eta_0^*$.

3. High gas contents, notably in biofoams, reveal a yield stress in liquids [56]. All these phenomena are also true for flows of gas-containing polymer melts. However, investigation of these flows has to account for normal stresses. In [57], Hee Ju Joo and C. D. Han made a theoretical and experimental analysis of the flow of a gas-containing polystyrene melt in merging (convergent) channels. The bubbles elongated along the flow axis, and normal stresses altered. It would seem that surface Laplace forces impede the formation of spherical bubbles into an anisodiametric body. The flow is further complicated by the spherical constituent of the stress tensor, which grows on approaching the narrow part in the convergent channel, resulting in partial, or sometimes even total dissolution of bubbles. The authors did not attempt to provide a complete mathematical description of the complex flow they observed. They emphasized in [57] that their theoretical analysis had a purely qualitative nature. Z. Shulman and S. Levitsky [58] examined, in their practice-oriented work, the dynamics of radial oscillations of a steam-gas bubble in a linear viscoelastic compressible inherited medium with reference to heat- and mass-exchange. The authors of [58] proposed the possibility of bubble oscillation with small losses in high-viscosity polymer fluids, for which a dissipation coefficient close to the initial medium viscosity was obtained with calculations of the Newtonian rheological model. Independently of [58], this effect was discovered in [59] for polybutadienes and polyethylenes. In other words, the introduction of gas bubbles into viscoelastic media with spectral relaxation properties alters both relaxation times and their distribution functions.

The above considerations illustrate the difficulties of trying to formulate equations descriptive of rheological behavior of polymer melts with gas bubbles. An optimistic approach to the solution of this task is contained in [60, 61]. The content of these works is revealed by their titles: On the Use of the Theory of Viscoelasticity for Describing of the Behaviour of Porous Material and for the Calculation of construction

Elements [60]; Technique for Determination of Rheological Characteristics of Porous Material on the Basis of the Theory of Viscoelasticity, and its Application for Calculation of Construction Elements [61]. The authors of [60, 61] theoretically and experimentally investigated foam polyurethane of the SISPUR SH4050/I brand. Their rheological equations were based on equations of viscoelasticity and integral-type equations. The nuclei in integral parts of equations that describe the relationship between deviatory and spherical constituents of stress tensors and strain tensors, were accepted as non-differential. The difference in the behaviour of material under positive and negative hydrostatic pressure is described with Heavyside functions. The equations presented in [60] are concretized in [61] and applied in the calculation of a trilayered construction element with a foam polyurethane middle layer.

Figure 1 in Ref. [60] represented creepage curves of the studied foam polyurethane sample, which show non-damping nature of creepage and the presence of viscous flow deformations on unloading. This determined the choice of nuclei in the studied equations when they were concretized in [61] in the form of power functions; this corresponds to non-damping ("indifferent") creepage. This means that the rheological formulas in [60, 61] describe the behaviour of what is really a viscoelastic *liquid* with gas bubbles. Naturally, even these complex formulas fail to account for numerous rheological effects which take place on flow of gas-containing polymer melts. However, we believe that the authors of [60, 61] formulated a fairly universal and very promising approach to a theoretical description of rheological behaviour of the media studied here; another argument in favour of their approach is the well-worn saying that any theory is nothing but a caricature of the real experiment.

It is often sufficient for the technologist to know the difference between the pressures, which is required to extrude the gas-liquid medium through the channel. This problem can be solved, at least for flows in straight pipes with an unchanging cross section, without resorting to rheological equations of such two-phase media. This idea is based on two concessions:

a) the flow is considered uni-dimensional; the channels with equal hydraulic radii are considered identical (cf. [62] for a review of simulation of flows in polymer processing with the hydraulic radius);

b) the liquid, per se, is considered non-compressible. Since the volumetric flow rate of liquid in each section along channel length is constant, the bubbles in the melt reduce the "live" section of the channel (i.e. area occupied by melt) resulting in greater apparent shift rates. In the first approximation, it is supposed that bubbles do not contribute to energy dissipation with their movement. With reference to concession a), the main formulas are as follows [63]:

$$S_L = S(1 - \varphi) \qquad (19)$$

where S_L is "live" channel section; S is total channel section, φ is instant volumetric content of free gas in section.

The ideas in a) and b) were used by D. H. Lewis and P. R. Ryason to describe flows of gas-containing polymer melts [64]. They considered gas insoluble in the melt. The effects of gas solubility are summed up in [20, 21].

From [20] and movement equations for unidimensional flow, we get

$$dP/dl \sim \eta(\dot{\gamma}_{app}) \cdot \dot{\gamma}_{app} \tag{20}$$

The authors of [20, 21] then supposed that gas bubbles in each section have time to reach equilibrium dimensions; then, at any pressure $P < P_{cr}$, φ in Eq. (19) can be determined from Eq. (9) and from putting φ in Eqs. (19) and (20). The influence of dissolved gas (its partial pressure ΔP_g) is assessed with Eq. (8) and Eqs. (15) and (16), with C_g from Eq. (8) substituted into Eq. (15) in place of the $(GC\frac{T_i}{T_0})$ complex. This makes it possible to draw up an equation for determination of pressure in any spot along channel axis when melt flows from point l_{cr}, which is determined by Eq. (17) (when $P_{ent} < P_{cr}$; $l_{cr} < 0$).

The simple result is attained when dependence of viscosity on shear rate $\eta(\dot{\gamma}_{app})$ can be ignored in the interval of apparent shear rates realized along channel axis. Then,

$$\eta(\dot{\gamma}_{app}) = \eta_{melt} \tag{21}$$

where η_{melt} is apparent viscosity of the pure melt realized on its flow with volumetric flow rate Q in the studied channel.

The approximated formula is as follows [20, 21]:

$$P = \frac{P_0 + \sqrt{P_0^2 + 4\left(1 + \frac{M}{f_{T_i}}\right)(K + M) P_0}}{2\left(1 + \frac{M}{f_{T_i}}\right)} \tag{22}$$

In Eq. (22), P_0 is pressure in the pure melt;

$$M = H_T \cdot \varrho_{melt}\ ; \qquad K = (GCT_i/T_0)\,(\varrho_{melt}/100 - c)$$

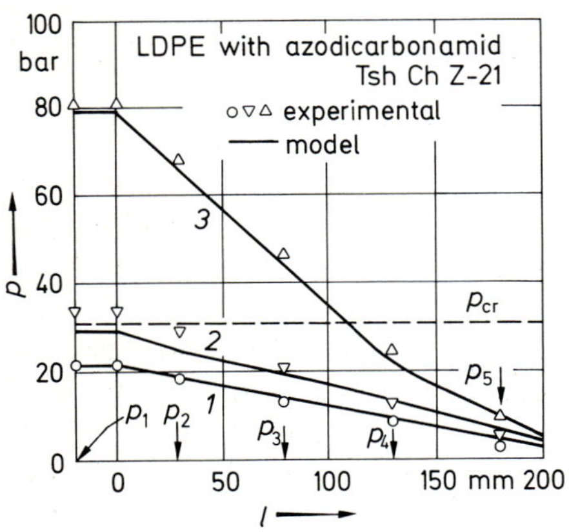

Fig. 5. Axial Pressure Profiles in Rectangular Slit Capillary (l = 200 mm; h = 2 mm; b — 10 mm); T = 190 °C; γ': 1.696 s^{-1} (*1*); 3.39 s^{-1} (*2*); 19.96 s^{-1} (*3*)

Equation (22) was obtained, essentially, with examination of the energy balance equation with respect to flows of gas-containing polymer melts. The key moment of this analysis is, in our view, comprehension of the fact that the energy of gas dissolved in the polymer is transformed into the energy of movement of the two-phase medium.

Figure 5 illustrates a fair correlation between theory and experiment.

6 Flow Structures

Let us return to Fig. 1 and its extrudate structures. The original photographs of extrudate sections are presented in [21]. Hydrodynamic characteristics of two-phase non-polymer systems and non-polymer foams have been theoretically analyzed by Soviet scientists — B. V. Deriagin [65], V. K. Tikhomirov [66], S. S. Kutateladze [67], R. I. Nigmatullin [68, 69], and may foreign scientists [70–72]. Models descriptive of two-phase flows, both vertical (ascending and descending) and horizontal, have been proposed [73–77]. S. S. Kutateladze and M. A. Styrikovich [67], along with G. Wallice [70], reported on three modes of flow for gas-liquid non-polymer mixtures in vertical straight pipes:

1. On low gas content and low speed of gas-water mixture, the bubbles are comparatively small and distributed throughout the section of the pipe in an even manner; this flow mode was termed "emulsion" or "bubble" flow;
2. Growing gas content makes bubbles merge until ultimately they occupy the central part of the pipe section, reaching 1 metre and more in length; these long bubbles are separated from the pipe wall by a thin layer of water. They have the shape of an ellipsoid at the front, and are "cut off" at the back, not unlike artillery shells. Hence the term, "shell" mode.
3. Further growth of gas content increases shell length; the quantities of water between them become smaller until finally, separate bubbles merge into a single entity — a bar surrounded by a ring of water: this is known as the "bar" mode.

Gravity acts as an additional influence in horizontal and sloping pipes, making flows stratify owing to lateral bubble migration. For gas-containing polymer melts, gravity can be ignored, since melts have high viscosities and, consequently, great Stokes forces of resistance to gas bubble motion by the liquid in the gravity force field.

As seen from Fig. 1, characteristics of gas-containing polymer melts reduced to channel dimensions can be divided into three areas depending on the type of extrudate obtained.

The first area covers low volumetric flow rates, and entrance pressures below P_{cr}. This sector of two-phase flow in the molding machine is characterized by a complex non-linear dependence of reduced pressure on reduced volumetric flow rate. The structure of foam plastics obtained in this way was called "shell" structure by the authors in [20, 21] — the extrudate contains huge shell bubbles which are comparable to its section. As CBA concentration increases, or medium volumetric flow rate is increased at low CBA concentration, small bubbles materialize in the melt around the shell bubbles, and the structure becomes "shell-bubble". Increase of the volumetric flow rate and the concentration of flowing agent neutralizes the difference in bubble size: their lateral dimensions become smaller than their longitudinal ones.

The second area is distinguished by that entrance pressures as a rule are equal to, or exceed, P_{cr}. (Q — P) dependences for gas-containing melts are shifted into the lower-pressure field to the constant ΔP parallelly to dependence for pure melt. This shift of Q — P characteristics is greater with greater CBA concentration, in accordance with the contribution of dissolved gas to overall pressure that is described by Eq. (15). This area is characterized by an even, fine-cell structure of obtained material; as reduced volumetric flow rate ($\dot{\gamma}'$) grows, there emerges a trend to stratification along the channel axis. Depending on CBA concentration (C), layer structures are reversed. For instance, if C = 0.1%, the small-cell structure core is surrounded by a ring flow of the polymer that contains no bubbles; if, on the other hand, C = 2 to 2.5%, the picture is reversed — the polymer rod is now surrounded with a ring of fine-cell material. Passing from low to high CBA concentrations is characterized with intermittent layers of fine-cell material and polymer with a few gas inclusions. The layered structure results from the formation of bubbles (i.e. gas desorption) primarily in the most advantageous spots, instead of migration in a direction normal to flow axis. This is confirmed by foaming of sections that contain no gas bubbles, on re-heating.

The reasons for such a layered structure are, in our view [20, 21], the lateral temperature gradients that are observed when the extrudate is cooled upon leaving the channel; and lateral pressure gradients. The resultant effect of both gradients determines layer order in the flow with intermittent gas content (0.5 to 1%). The lateral pressure gradient is conditioned by normal stresses in the flow. This was discussed in [8, 9].

Let us examine the influence of the temperature gradient on extrudate structure.

Investigation of the movement of gas inclusions in liquids and solids under the influence of the temperature gradient is of interest in many domains of science and technology. For example, the gas bubbles that form in the weld seam in metal processing, detract from its strength. This problem is highly important, and it was not for nothing that Soviet cosmonauts performed welding operations in zero gravity, which removed any possibility of a gravity force field and presented the effect of the temperature gradient on the structure of the seam, notably, on the amount and location of gas inclusions therein. Those interested in a purely scientific analysis can address themselves to [78], which examines movement of inclusions in solids under influence of the temperature gradient and non-homogeneous stress field. We shall limit ourselves to a qualitative discussion of this problem, with the support of Prof. Ya. E. Geguzin's popular work, "Bubbles" [27]:

a) if a temperature gradient is at work in a liquid, the bubbles are affected by the driving force

$$F = -\pi r^2 \cdot d\sigma/dT \cdot dT/dx \tag{23}$$

where r is bubble radius, and σ is surface tension. As seen from Eq. (23), the bubble moves towards heat in a non-uniformly heated liquid. This is particularly important at the liquid-solid interface. Gas solubility in crystals is orders lower than in liquids. The hardened polymer "ejects" gas, and the formed bubbles move from the crystallization front towards hotter liquid.

b) gas thereby accumulates in advance of the front of crystallization. Formation of bubbles can occur only after gas concentration in the vicinity of the solid-liquid interface reaches a certain critical point, i.e. a concentration at which pressure in the polymer attains P_{cr}. The liquid at the interface then starts to "boil": a layer of bubbles appears and, provided that the front of crystallization moves sufficiently rapidly, is engulfed by the hardened polymer. After that, the front of crystallization has to make some more headway for concentration of gas in the liquid to grow critical again. Gas bubbles are captured by the moving front of crystallization in a periodic way! In this case, as Fig. 1 shows, extrudate structure along the radius takes the shape of alternating foamed and non-foamed polymer rings. Similar phenomena were observed upon water crystallization; p. 145 of Ref. [27] presents photographs of ice structures with captured gas bubbles, that are analogous to those in Fig. 1.

Now back to Refs. [7, 20, 21] and Fig. 1.

The third area of P — Q characteristics corresponds to the onset of irregular flow (LDPE) or stick-slip (linear HDPE — for instance, Lupolen by BASF). For LDPE, reduced (Q — P) characteristics approach, with the growth of $\dot{\gamma}'$ and C, corresponding curves for pure polymer melt. For HDPE, stick-slip with growth CBA concentration degenerates, and the dependences acquire a S-shaped form. As for the structures obtained in modes corresponding to this area of the curves, they are irregular at all CBA concentrations. However, layers rotate in this area, too. For instance, at C = (0.1–0.5)%, the screw-distorted or chaotically disrupted rod of fine-cell material is surrounded by the polymer which contains no bubbles; the surface of the extrudate is smooth despite considerable distortions of the form of the flow itself. At $C \geqq 1\%$, on the other hand, the fine-cell flow with large surface deformations includes a screw-shaped bar or its fragments consisting of a polymer with solitary gas bubbles.

It should be pointed out that the third area corresponds to pressures at the entrance that exceed the critical point P_{cr} and, consequently, the pressure corresponding to the transition towards the irregular flow mode, with reference to the foregoing, equals:

$$P^{irr}_{foam} = P^{irr}_{0} - \Delta P \tag{24}$$

In Eq. (24), P^{irr}_0 is pressure at which transition to the irregular flow mode of pure melt is attained; ΔP is calculated from Eq. (15). Equation (24) was experimentally tested in [21] for the same samples and in the same channels as were used for (15); cf. Table 1 for results of comparison of Eq. (15) with experiments. Deviation between P^{irr}_{foam} calc and P^{irr}_{foam} exp is of the same order as ΔP_{calc} and ΔP_{exp} in Table 1.

An interesting and practically valuable result was obtained in [21] for PE + N_2 melts, and in [43] for PS + N_2 melts. The authors classified upper critical volumetric flow rate and pressure with reference to channel dimensions: $\tau' \sim P^{irr}_{fe\,im}$; $\dot{\gamma}' \sim Q^{irr}_{fe\,im}$. Depending on volume gas content φ at channel entrance (pressure of 1 stm., experimental temperature), τ' and $\dot{\gamma}'$ fall, in accordance with Eq. (24), to $\varphi \approx 0.85$. At $\varphi \approx 0.80$, in a very narrow interval of gas concentrations, τ' and $\dot{\gamma}'$ fall by several orders. The area of bubble flow is removed entirely. It appears that at this concentration of free gas, a phase reversal takes place as the polymer melt ceases to be a continuous phase (fails to form a "continuous cluster", in flow theory terminology). The theoretical value of the critical concentration at which the continuous cluster is formed equals 16 vol. % (cf., for instance, Table 9.1 in [79] and [80]). An important practical conclusion ensues; it is impossible to obtain extrudate with over 80% of cells without special techniques. In other words, technology should be based on a volume concentration of gas that would not exceed 80% in any spot of the channel. Flow stick-slip degenerates on growing gas concentration in HDPE [7], just as it does when a low-molecular solvent is injected into a polymer [50]. When azodicarbonamide attains a concentration of 1% (by weight), irregularity of flow is observed for Lupolen-type HDPE; simultaneously, the sector of rapid $\dot{\gamma}'$ growth in the narrow τ' interval, which is characteristics of flow stick-slip, disappears from the reduced P — Q characteristics. However, the analogy with concentrated polymer solutions is not total. Where critical rates grow as the latter are diluted, they drop with growing gas in gas-filled polymers. For nitrogen-filled PE melts [20, 21] and for PS + N_2 [26, 43], a "lowermost limit" has been established for bubble flow modes. It has the following form:

$$\lg(P_{ent}/P_{cr} \cdot \dot{\gamma}') = -E - U \lg(4\sqrt{S}/l) \qquad (25)$$

In Eq. (25), P_{ent} is the entrance pressure of the gas-containing composition; S is the cross-section area; l is its length, E and U are constants. Other variables are marked as before.

Using the above formulas, it is possible to calculate the values of formulas under the log sign in the left- and right-band parts of Eq. (25), and to chart dependence (25) theoretically, Fig. 6 illustrates fair correlation of theory and experiment.

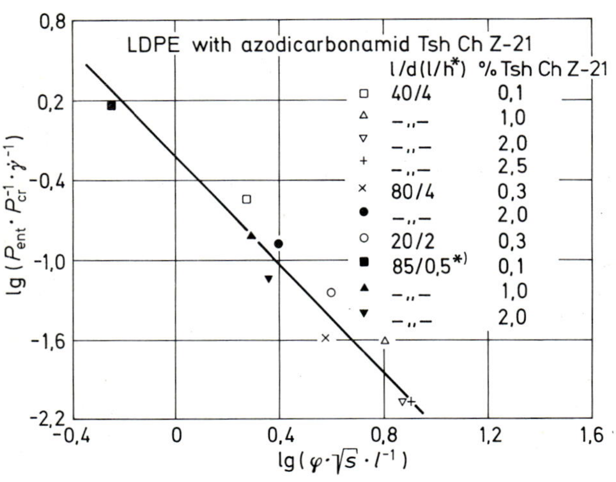

Fig. 6. Generalized Description of Lower Border of the Straight Section (with bubble structure) of Reduced Q — P Characteristics of Composition Flow (compositions based on LDPE, 190 °C, with varying CXZ-21 azodicarbonamide content)

7 Conclusion

The hydromechanics of flows of gas-containing polymer melts is extremely complicated to analyse and describe mathematically. Nevertheless, comprehensive experimental investigations in this sphere have yielded simple models of unidimensional flows, and solved several problems pertaining to the physics and hydromechanics of complex medium flows.

Recently elaborated methods for predicting volumetric flow rate/pressure characteristics, extrudate structures, and determining high and low limits of optimal processing, can now be used in industrial processing of plastics.

These methods rely on temperature changes, alternation of channel dimensions and flow rates, to produce foamed articles of a required structure, which is important for manufacturing foamed and foam-filled articles made of plastics, as well as for equipment design in this sphere.

We would like to think that this work will stimulate theoretical and experimental interest in this sphere of rheology, which is vitally important for polymer processing.

8 References

1. Berlin AA, Shutov FA (1980) Chemistry and technology of gas-filled high polymers (in Russian). Nauka Publ, Moscow
2. Han CD, Ma CJ (1983) J Appl Polym Sci 28: 831
3. Han CD, Ma CJ (1983) J Appl Polym Sci 28: 851
4. Han CD, Ma CJ (1983) J Appl Polym Sci 28: 2961
5. Han CD, Ma CJ (1983) J Appl Polym Sci 28: 2983
6. Yang HH, Han CD (1985) J Appl Polym Sci 30: 3297
7. Nikolaeva NE, Sabsai O Yu, Fridman ML (1984) Doklady AN SSSR. 276: 140
8. Han CD, Villamizar CA (1978) Polym Eng. and Sci 18: 687
9. Villamizar CA, Han CD (1978) Polym Eng and Sci 18: 699
10. Piperov NL (1976) Plastich massy 5: 38
11. Han CD, Kim JW, Malhotra KD (1976) Journ of Appl Polym Sci 20: 1583
12. Lundberg JL, Wilk MB, Huyett MJ (1962) Journal of Appl Physics 31: 1131
13. Lundberg JL, Wilk MB, Huyett MJ (1962) Journ of Appl Polym Sci 57: 275
14. Durril PL, Griskèy RG (1966) AIChE Journal 12: 1147
15. Durril PL, Griskey RG (1969) AIChE Journal 15: 106
16. Strikman SW (1974) Rubber Age, p 58
17. Hobbs SJ (1975) Polym Eng and Sci 15: 854
18. Saunders JH, Hansen RH (1972) Plastic Foams (Pt. 1). M. Decker, New York, p 39
19. Bikerman JJ (1973) Foams Springer, Berlin Heidelberg New York
20. Nikolaeva NE, Sabsai OYu, Fridman ML (1985) Rheological properties of gas-filled polymers and their effects on processing Ser. Pererabotka plastmass. NIITEHIM, Moscow
21. Nikolaeva NE (1984) Flow of gas-containing polyethylene melts on extrusion molding. Thesis (in Russian). MITHT, Moscow
22. Chemist's Manual (in Russian) (1964) Khimiya Publ Moscow-Leningrad 3: 1008
23. Avdeev NYa (1967) Koll Zhurnal 29: 315
24. Volarovich MP, Avdeev NYa (1964) Koll. zhurnal. 26: 647
25. Yang HH, Han CD (1984) J Appl Polym Sci 29: 4465
26. Kascheeva NI, Mitrofanov AD, Mamontov VM, Andreevskaya LV (1987) Plast massy 3: 57
27. Geguzin EYa (1985) Bubbles (in Russian). Nauka Publ., Moscow
28. Han CD, Hee Yu Joo (1981) Polym Eng and Sci 21: 518
29. Frenkel YaI (1946) Journ of Experim and Theor Physics 16: 29
30. Hee JuJoo, Han CD (1982) Amer Inst Chem Eng. Journal 28: 1002
31. Han CD (1979) Rheology in polymer proceessing (in Russian) Khimia Publ Moscow
32. Astarita G, Marrucci G (1974) Principles of non-Newtonian fluid mechanics. McCrow-Hill
33. Amon M, Denson CD (1984) Polym Eng and Sci IA 24: 1026
34. Amon M, Denson CD (1986) Polym Eng. and Sci. IIA 26: 255
35. Sosin SE (1984) Summary of Cand Thesis INHS AN SSSR, Moscow
36. Vinogradov GV, Elkin AJ, Sosin SE (1978) Polymer 19: 1458
37. Cole R (1974) Adv. in Heat Transfer 10: 85
38. Blander M, Katz JL (1975) AIChE Journal 21: 833

39. Blyler LL, Kwei TK (1971) Journ of Polym Sci 35: 165
40. Kuharsky YuM, Mitrofanov AD, Kozlov NA, Kudriashov VP (1987) Plstmassy 3: 16
41. Okunev PA, Panov YuT, Mitrofanov AD (1982) Ser Khimia i Khimicheskaya technologia 25: 477
42. Oyanagi J, White JL (1979) Journ of Appl Polym Sci 23: 1013
43. Kascheeva NI, Mitrofanov VD, Sabsai OYu, Nikolaeva NE (1988) Izvestia vuzov. Khimia i Khimicheskaya technologia
44. Crank G, Park GJ (1968) Diffusion in polymers. chapt. 2. Academic Press, London
45. Haward RN (1970) J Macromol Sci 2: 191
46. Privalko VP (1986) Molecular structure and properties of polymers (in Russian) Khimia Publ, Leningrad
47. Bueche F (1962) J of Chemical physics 36: 2940
48. Maklakov AI, Skirde VD, Fatkullin NF (1987) Self-diffusion in polymer solutions and melts (in Russian). University Publ, Kazan
49. Kimmich R (1985) Helv phys acta 58: 102
50. Vinogradov GV, Malkin AYa (1977) Polymer rheology (in Russian) Khimia Publ, Moscow, pp 123, 222
51. Bolotov AA, Mirzadzhanzade AA, Nesterov II (1988) Mekh. zhidk i gaza 1: 172
52. Kraynik AM (1981) Polym Eng and Sci 21: 80
53. Nigmatullin RI (1987) Dynamics of multiphase media (Two Parts, in Russian) Nauka Publ, Moscow
54. Bachelor G (1973) Introduction to fluid dynamics (in Russian) Mir Publ, Moscow
55. Davies R (1954) Proceedings of the Royal Society of London. A226: 34
56. Wenzel HG, Brungaber RY, Stelson TE (1970) Journ of Materials 5: 2
57. Hee Ju Joo, Han CD (1981) Polym Eng and Sci 21: 69
58. Shulman ZP, Levitsky SP (1980) Doklady AN SSSR. XXIV: 1088
59. Fridman ML, Peshkovsky SL, Tukachinsky AI, Vinogradov GV, Eenikolopyan NS (1983) Polymer Composites 4: 126
60. Pozdeev AA, Chuvatova IV (1982) In: Stress-deformed state and strength of constructions (in Russian), UNTc AN SSSR, Sverdlovsk, p 31
61. Chuvatova IV (1982) In: Stress-deformed state and stength of constructions (in Russian), UNTc AN SSSR, Sverdlovsk, p 36
62. Tin C (1985) In: Developments in plastics technology. Elsevere Applied Sci. 2: 297
63. Kutateladze SS, Styrikovich MA (1958) Hydraulics of gas-liquid systems (in Russian). Gosenergoizdat Publ., Moscow-Leningrad
64. Lewis DH, Rayson PR (1980) J Appl Polym Sci 25: 1695
65. Deryagin BV, Titievskaya AS (1953) Koll zhurnal 15: 416
66. Tikhomirov VK (1975) Foams, theory and practice of their forming and coalescence (in Russian). Khimia Publ, Moscow
67. Kutateladze SS, Styrikovich MA (1958) Hydraulics of gas-liquid systems (in Russian). Gosenergoizdat Publ, Moscow
68. Nigmatullin RI (1978) Basic mechanics of heterogeneous media (in Russian). Nauka Publ, Moscow
69. Nigmatullin RI, Habeev NS (1976) Izvestia AN SSSR. Mekh. zhidk. i gaza. 6: 56
70. Wallice G (1972) Uni-dimensional two-phase flows (in Russian). Mir Publ, Moscow
71. Hewitt GF (1983) Int. J. Multiphase Flow. 9: 715
72. Martin CJ, Whalley PB (1983) Int Conf Phys Modell Multiphase Flow Coventry Apr, Cranfield, p 283
73. Nesterovich NI (1979) Equations of turbulent motion of heterogeneous mixtures (in Russian). Prepr of Inst of Theor and Appl Mech USSR Academ of Science, Siberian Branch 8: 28
74. Chen JJJ, Spedding PL (1983) Proc Condens Pap Miami Beach FL 18 20 Apr Coral Babbles, Fl p 15
75. Shevelev AV, Milkov VA, Abayev GN (1979) Manuscript No 2699/79 deposited with ONIITEHIM, Tcherkassy, May 14. Polytechn Inst, Yaroslavl
76. Chisholm D (1983) Int Conf Phys Modell Multiphase Flow Coventry Apr, Cranfield, p 219
77. Choudhury D Sen, Radhakrishnan VR, Mitra AK (1983) Phys Modell Multiphase Flow Coventry, Apr Cranfield, p 255

78. Geguzin Ya E, Krivoglaz MA (1971) Movement of macroscopic inclusions in solids (in Russian). Metallurgic Publ, Moscow
79. Ziman YM (1979) Models of disorder. Cambridge University Press, Cambridge
80. Stanffer D (1985) Introduction to percolation theory, London

Editor N. S. Enikolopian
Received July 27, 1989

Formation Processes and Properties of Conducting Polymer Composites

A. T. Ponomarenko, V. G. Shevchenko, N. S. Enikolopyan
Institute of Synthetic Polymer Materials of the USSR Academy of Sciences, Moscow

Conducting polymer composites (CPCM) have usually been considered from the point of view of their constituents and properties. However, processing conditions may influence the properties even more strongly. The present article is meant to redress the situation, relating processing conditions and the final properties of these materials. First, components, structure, and conductivity of CPCM are shortly reviewed, with the newly developed types of conducting fillers included. Then, redistribution of disperse filler particles during injection molding and resulting conductivity profiles are considered. The use of polymer blends may lead to the segregation of the filler, resulting in higher conductivity at lower concentrations. These phenomena are also considered. The next chapter deals with the influence of processing on conducting fibers, which are the most promising fillers but which easily break under extreme processing conditions. A novel technique for manufacturing CPCM — polymerization filling — and the properties of resulting composites are reviewed. Finally, existing and future applications of conducting polymer composites are considered.

1 Introduction 126

2 Constituents of Conducting Polymer Composite Materials 127

3 Structure and Conductivity of CPCM 128

4 Processing Conditions and Conductivity of Composite Materials 131
4.1 Redistribution of Filler Particles During Processing 133
4.2 Profile of Conductivity in the Process of Injection Molding 134
4.3 Distribution of Filler in Conducting Polymer Composites Based on Polymer Blends 136
4.4 Some Features of Processing of Composites with Conducting Fibers . . 138
4.5 Polymerization Filling — A Novel Technique for Manufacturing Conducting Composite Materials 140

5 Applications of Conducting Polymer Composite Materials 142

6 Summary 145

7 References 145

Advances in Polymer Science 96

1 Introduction

Conducting polymer composite materials (CPCM) — artificial media based on polymers and conductive fillers, have been known since the early 1940s and widely used in various branches of science and technology. Their properties are described in a considerable number of monographs and articles [1–12]. However, the publications available do not clearly distinguish such materials from other composites and do not provide for specific features of their formation.

The main feature of the CPCM is a drastic difference between electric conductivity of a polymer matrix and the filler reaching a factor of 10^{24} in terms of resistivity (Fig. 1). There is no such difference in relation to any other physical property of com-

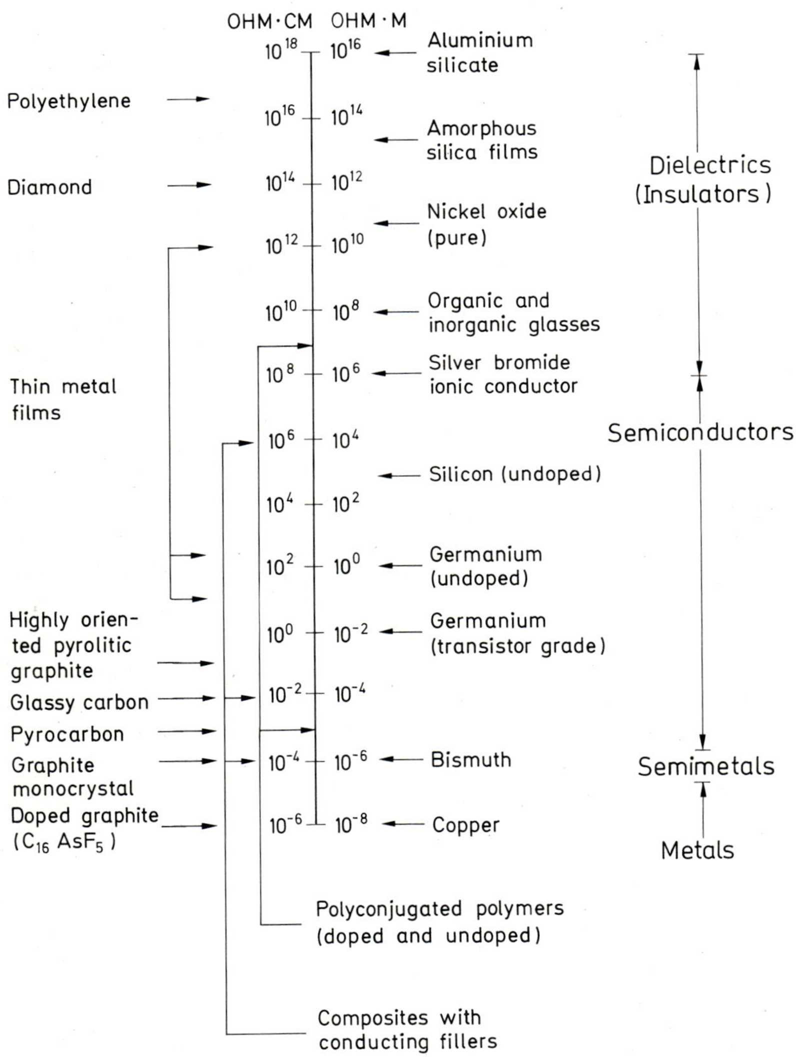

Fig. 1. Conductivity levels of various materials

posite constituents. This is due to the important specific feature — direct current can flow along such materials only through continuous chains of filler particles (if the possibility of quantum-mechanical tunnelling is not taken into account). Because of the small (as compared with the sample volume) size of filler particles, excluding conducting textures, there is a great number of variants of their mutual arrangement, many of which are indistinguishable, however, only a small portion will satisfy the spatial structure where current flow is possible. The probability of forming such a structure depends on a variety of factors, namely: quantity of a filler, shape of filler particles and their capability of becoming combined with a polymer, method of mixing, etc. The influence of many factors is clear at present but is not always considered in creation of particular composites. In the present review the authors try to describe the sequence of a designing process of a composite material possessing conducting properties, beginning with a choice of constituents and ending with a method of their combination. The authors hope that the review will be useful not only to research workers but also for practical personnel engaged in a given branch of polymer materials technology.

2 Constituents of Conducting Polymer Composite Materials

The CPCM constituents are a conducting filler and polymer matrix where this filler is dispersed randomly.

Almost any known polymer or polymer mixture can be used in the capacity of a polymer matrix; various additives may be introduced in the matrix to reduce melt viscosity, increase thermal stability of the composition or its plasticity, etc. A choice of a matrix is determined mainly by the operating conditions of a material and the desired physical-mechanical properties of a composite. One may state rather confidently that, other things being equal, the value of the CPCM conductivity does not depend on a choice of a polymer matrix [3].

When selecting a polymer-filler pair, one should take into account their compatibility which, to a first approximation, is determined by the parameters of adhesive interaction U_{ff} (interaction of filler particles) and U_{fp} (interaction of a filler and a polymer). Let $U_{ff} > U_{fp}$, then filler particles interacting mostly with each other stick together and form aggregates which prevent the formation of current-conducting chains. Otherwise, with $U_{ff} < U_{fp}$, the filler particles will be coated with a layer of a polymer due to good adhesion of the polymer to the filler, and remain electrically isolated, and this will also prevent the formation of the current-conducting chains. Apparently, the best results can be achieved in the intermediate case when the surface of the filler particles has zones with different adhesion properties, the part of which provides for the formation of a chain structure while the others ensure interaction with the polymer matrix. In practice, these magnitudes may vary by modifying the filler surface, introducing anti-adhesive agents, changing temperature and other conditions.

As far as conducting fillers are concerned, we have rather a wide range of choice. In addition to the traditional and long used fillers, such as carbon black and metal powders [13] fiber and flaky fillers on organic or metal bases, conducting textures, etc recently appeared and came into use. The shape of the filler particles varies widely, but only the particle aspect ratio, the main parameter which determines the probability of neighboring particle contact at a given degree of filling, is important from the viewpoint of using them in the CPCM. The aspect ratio l/d is a ratio of maximum (l) and minimum (d) dimensions of a geometrical body, and from this point of view, a sphere, cube, ellipsoid are approximately similar bodies having $l/d \approx 1$.

Based on the above facts, it is convenient to classify the conducting fillers according to the following parameters:

1 — material (organic, metallic);
2 — aspect ratio l/d ($l/d \approx 1$ — disperse; $1 < l/d < 1000$ — flakes; short fibers; $l/d > 1000$ — long fibers);
3 — distribution of filler particles according to size — F (l, d, l/d);
4 — Filler conductivity.

The properties of traditional fillers, such as carbon black, graphite, metal powders, carbon fibers, are described in detail in [13], therefore, new kinds of conducting fillers which have recently appeared will be considered below.

At present, the most promising fillers are those with $l/d \gg 1$, i.e. fibers and flaky fillers that make it possible to reduce filler concentration in a composite and, thus, facilitate the processing and improve physical-mechanical properties [17]. Besides cut carbon fibers, carbon fibers coated with a layer of Ni that have higher conductivity have been developed (American cyanamid) [14]. Glass fibers with a layer of aluminium (MB Associates, Lundy Electronics) [16] are in production.

But carbon and glass fibers are notable for their fragility and are inconvenient for processing and, therefore metal fibers of brass (Aron Kasei Co.), copper (Toshiba Chemical Corp.) [14], stainless steel (Brunswick Corp., Bekaert) [16], aluminium (MB ASSOCIATES, TRANSMET) [15] have been introduced.

Stainless steel fibers have attracted widespread attention. They are manufactured with diameters from 4 to 25 μm and lengths from 3 to 8 mm.

The fibers possess sufficient strength which makes it possible to maintain fiber aspect ratio under extreme conditions of processing (extrusion, injection moulding). The fibers are produced not as a powder but in bundles containing several thousand elementary fibers fixed together with a polymer binder (6 to 8 per cent by weight) which should be compatible with a polymer matrix of a proposed composite. Such bundles can be directly introduced into the melt zone of an extruder and this precludes a mixing stage and reduces fiber destruction during processing. Fiber resistivity is about 75×10^{-6} Ohm cm.

Among other fillers of note are aluminium flakes [15], hybrid fillers — a mixture of flakes and fibers (Transmet Corp.) [14], mica coated with a layer of Ni (Martin Marietta) [14], etc.

3 Structure and Conductivity of CPCM

Conducting polymer composite materials are typical disordered structures consisting of randomly (or according to a certain law) arranged particles of a conducting filler that are submerged into a polymer medium. In this case the filler particles have macroscopic (as compared with a free path of electrons in the particles) dimensions which facilitates the study of such structures allowing one to neglect quantum effects. However, the role of such effects in the CPCM is practically unstudied, though in some cases, say, at the filler-polymer boundary it may prove to be significant (e.g. the fluctuation tunnelling model allowing for quantum modulation of carrier transfer between conducting particles at low temperatures [2, 18]. One more simplification, which is not stated but in most cases is tacitly assumed, relates to the role of a polymer matrix which is considered as a certain continuous medium with dielectric constant ε′ that serves only to fix filler particles in space. Such an approach is useful for the consideration of the CPCM general properties and true, with a sufficient degree of accuracy, for amorphous polymers. However, it is not satisfactory in the case of, say, crystallizing polymers as the filler where particles are pushed out of crystal regions because of their large size and become concentrated in amorphous zones. This means inhomogeneous spatial distribution of the particles, or, in other words, their segregation [19].

The percolation theory [5, 20–23] is the most adequate for the description of an abstract model of the CPCM. As the majority of polymers are typical insulators, the probability of transfer of current carriers between two conductive points isolated from each other by an interlayer of the polymer decreases exponentially with the growth of gap l_g (the tunnel effect) and is other than zero only for $l_g < 100$ Å. For this reason, the transfer of current through macroscopic (compared to the sample size) distances can be effected via the contacting-particles chains. Calculation of the probability of the formation of such chains is the subject of the percolation theory. It should be noted that the concept of contact is not just for the particles in direct contact with each other but, apparently, implies convergence of the particles to distances at which the probability of transfer of current carriers between them becomes other than zero.

The main notion of the percolation theory is the so-called percolation threshold C_p — minimal concentration of conducting particles C at which a continuous conducting chain of macroscopic length appears in the system. To determine this magnitude the Monte-Carlo method or the calculation of expansion coefficients of C_p by powers of C is used for different lattices in the knots of which the conducting par-

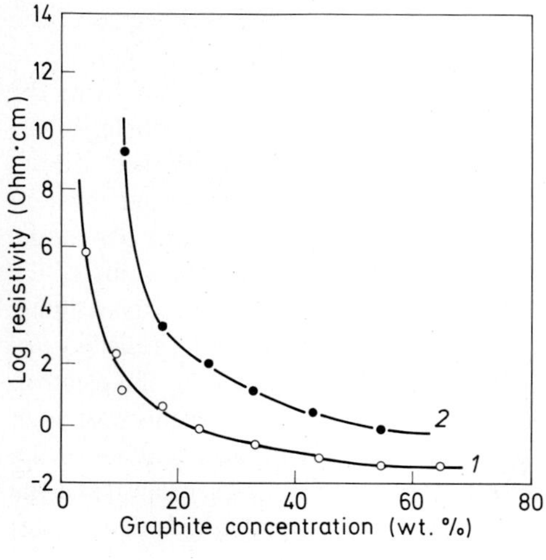

Fig. 2. Relationship of conductivity of polypropylene-based polymer composites and filler concentration (natural graphite): *1* — polymerization filling; *2* — mechanical mixture [24]

ticles are located. The value of the percolation threshold for a continuous three-dimensional medium equals $C_p \sim 0.15$ [29]. The detailed calculations, approximate dependencies of conductivity on concentration for different models and other details can be found in Refs. [5, 22, 29].

Experimental dependences of conductivity σ of the CPCM on conducting filler concentration have, as a rule, the form predicted by the percolation theory (Fig. 2, [24]). With small values of C, σ of the composite is close to the conductivity of a pure polymer. In the threshold concentration region when a macroscopic conducting chain appears for the first time, the conductivity of a composite material (CM) drastically rises (resistivity ϱ_v drops sharply) and then slowly increases practically according to the linear law due to an increase in the number of conducting chains.

Calculation of dependence of σ on the conducting filler concentration is a very complicated multifactor problem, as the result depends primarily on the shape of the filler particles and their distribution in a polymer matrix. According to the nature of distribution of the constituents, the composites can be divided into matrix, statistical and structurized systems [25]. In matrix systems, one of the phases is continuous for any filler concentration. In statistical systems, constituents are spread at random and do not form regular structures. In structurized systems, constituents form chain-like, flat or three-dimensional structures.

The maximum values of the percolation threshold are characteristic of matrix systems in which the filler does not form the chain-like structures till large concentrations are obtained. In practice, statistical or structurized systems are apparently preferable because they become conductive at considerably smaller concentrations of the filler. The deviation of the percolation threshold from the values of C_p to either side for a statistical system (~ 0.15) can be used to judge the nature of filler distribution.

An example of a structurized system is a composite obtained by pressing a mixture of the polymer powders and metal with radii of R_p and R_m, respectively ($R_p > R_m$) [19, 26]. When the powders are being mixed, metal particles stick around polymer granules and, if the pressing temperature is lower than the polymer melting temperature, do not penetrate inside the granules. As the material volume accessible for the filler is thus artificially reduced, conducting chains are formed at lesser concentration and the larger the ratio R_p/R_m the more reduced is the percolation threshold.

The composites with the conducting fibers may also be considered as the structurized systems in their way. The fiber with diameter d and length l may be imagined as a chain of contacting spheres with diameter d and chain length l. Thus, comparing the composites with dispersed and fiber fillers, we may say that $N = l/d$ particles of the dispersed filler are as if combined in a chain. From this qualitative analysis it follows that the lower the percolation threshold for the fiber composites the larger must be the value of l/d. This conclusion is confirmed both by the calculations for model systems [27] and by the experimental data [8, 15]. So, for $l/d \sim 10^3$ the value of the threshold concentration can be reduced to between 0.1 and 0.3 per cent of the volume.

As was noted above, when the conducting filler particles converge to a distance of $l_g < 100$ Å, the current transfer among them is possible by tunnelling the electron through a quantum-mechanical barrier. However, the equivalent resistance of such a junction is rather high, in any case it is higher than that with direct contact of the particles. These places may be considered as defects in the chain-like structure of the composite. The defects of other kinds may be associated with the high contact resistance in the places where the particles touch due to oxide layers or other admixtures present on their surfaces. Whatever nature of the defect may be, the presence of the defects makes the introduction of an extra amount of the filler necessary to compensate for the defect. This is not always desirable for various reasons. But such defects may be cured by the application of a high dc voltage which results in the electrical breakdown of the insulating interlayers [28]. As was proved, for $C < C_p$ the effect of 600 V pulses with a duration of 1 ms is absent in the compositions made up of polytetrafluoroethylene and carbon fibers. With the initial value of ϱ_v^0 of the composition being 1.8×10^4 Ohm × cm, after the effect of the pulse action ϱ_v decreases 37 times, while with $\varrho_v^0 = 0.92$ Ohm · cm, it decreases only 1.4 times. Thus, the smaller the difference in $C—C_p$, the higher is the effectivity of curing the defects by the action of high voltages [29].

The defects caused by the high contact resistance especially manifest themselves in the metal-filled composites where the value of the percolation threshold may reach 0.5 to 0.6 [30]. This is caused by the oxidation of the metal particles in the process of CPCM manufacture. For this reason, only noble metals Ag and Au, and, to a lesser extent, Ni are suitable for the use as fillers for highly conductive cements used in the production of radioelectronic equipment [32].

The CPCM structure also determines the following properties important in practice: the temperature coefficient of resistance, dependence of conductivity on frequency, etc. However, the scope of this review does not include the consideration of such dependences and they can be found in [2, 3, 12].

We conclude that the preparation of the samples of the polymer composites with the corresponding electrical properties in the form, say, of the plates, bars, hollow cylinders, etc., that are usually used for the purpose of research in the laboratories, and of real articles should be considered as two interrelated problems. This is important and should be stressed, as the values of the conductivity and other parameters obtained for the simple forms might prove different for the forms that may be used as constructional elements. Therefore, these circumstances should be taken into account at the design stage of a conducting composite as well as the optimum technological techniques of molding of practically important articles.

4 Processing Conditions and Conductivity of Composite Materials

The wide use of articles made of conducting composite materials depends on reproducibility of their physical-mechanical and electrical properties. Consistent mechanical properties (strength, modulus of elasticity) are comparatively easy to obtain, however the task of obtaining stability of electrical properties is rather complicated. One can even encounter statements in the literature that the production of conducting composite materials is linked with "black magic" [14]. It is bound up with the fact that mechanical properties of composite materials are mainly determined by average filler concentration, whereas electrical properties depend on whether a continuous net of contacting filler particles is formed in the material, i.e. on their distribution and local concentration. In other words, in technological chain "constituents → mixing → CCM → processing to article → article" each step makes its own, frequently unpredictable, contribution to a change of local concentration of the filler.

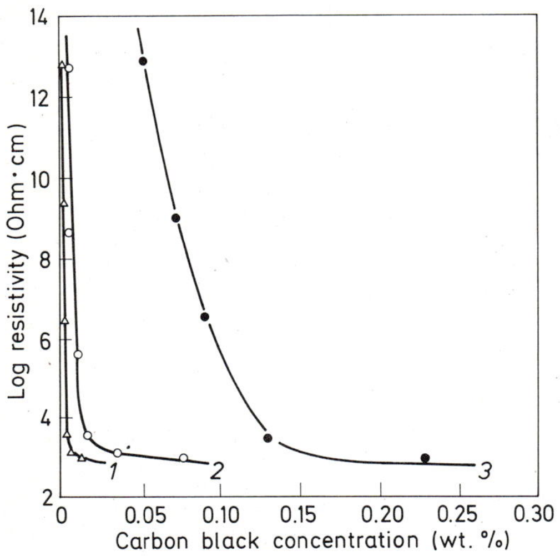

Fig. 3. Conductivity-concentration relationship of composite depending on manufacture technique [33]

The above may be illustrated by the following significant example [33]. Figure 3 shows the relationship of conductivity of polystyrene-based composite materials and the filler concentration (carbon black Ketjen-black EC). The materials are fabricated by different methods: (1) pressing of a mixture of carbon black and polymer powder; (2) mixing of carbon black with toluene-dissolved polymer in a ball mill, moulding of film by slow evaporation of the solvent, grinding of the film and pressing; (3) calendering of a mixture of carbon black and polymer between heated rollers, grinding of the product and pressing.

In pressing, the threshold concentration of the filler amounts to about 0.5% of volume. The resulting distribution of the filler corresponds, apparently, to the model of mixing of spherical particles of the polymer (with radius R_p) and filler (with radius R_m) for $R_p \gg R_m$ as the size of carbon black particles is usually about 1000 Å [19]. During this mixing, the filler, because of electrostatical interaction, is distributed mainly on the surface of polymer particles which facilitates the forming of conducting chains and entails low values of the percolation threshold.

For the second method the threshold concentration of the filler in a composite material amounts to about 5 volume %, i.e. below the percolation threshold for statistical mixtures. It is bound up with the fact that carbon black particles are capable (in terms of energy) of being used to form conducting chain structures, because of the availability of functional groups on their surfaces. This relatively "sparing" method of composite material manufacture like film moulding by solvent evaporation facilitates the forming of chain structures.

During calendering, the value of threshold concentration of the filler rises to about 25% by volume, i.e. above the threshold for statistical mixtures. Obviously, in this case mechanical forces, especially shearing stresses, destroy the forming chain structures of black particles. The data on the effect of prolonged mixing in an extruder on the composite material conductivity are indicative of the same fact. In the case of thorough mixing, the value of conductivity drops sharply due to the break up of soot aggregates and uniform distribution of the carbon black in the composite material. These data were obtained in researching the mixtures of carbon black and copolymer of propylene and vinyl chloride [34], and the mixture of high-density polyethylene copolymers of ethylene with ethyl acrylate and vinyl acetate [35].

The above data provide the proof of the crucial effect of a method of constituents mixing and composite material processing on the distribution of the filler and, thus, on electrical properties of conducting composite materials. These issues are considered hereinafter.

4.1 Redistribution of Filler Particles During Processing

Polymer processing has been studied extensively, both from theoretical and practical viewpoints (See, for example, [36]). However, the processing of polymers with fillers is considered usually from the "local" viewpoint as the processes of dispersion, i.e. the criteria for the existence of shearing stresses, that separate particle pairs or clusters in the polymer melt flux, are sought. In this case the researchers ignore the fact that the final distribution of the filler in the material or article might prove, to a great extent, to be nonuniform, which would affect physical-mechanical and moreover electrical properties of the composite material. Some aspects of this problem are considered in the works by F. Gauthier and his collegues [37, 38], by P. Brunn [39] and by S. Emerman [40] as well as during the investigation of model systems [41, 42].

The work [40] deals with the redistribution of filler particles in the process of injection molding. In this case nonuniform distribution may occur both in the cross-section of a sample and along its length. Both kinds of nonuniformity are linked together: if particle moves away from the mold walls it enters the zone of high velocity flow, therefore, a deficit of particles near the walls should be accompanied with a surplus of them far from the inlet. It should be noted that all the works mentioned consider spherical particles: there are no theoretical or experimental studies of the redistribution of particles of other shapes, say, fibers or bars.

From dimension considerations it follows that the speed of radial migration of particles (laminar flow in a cylindrical tube is under study) depends on the following parameters: r — a distance along the radius from tube centre; R — tube radius; a — radius of a spherical particle of filler; v — volumetric flux of a polymer, n — index of a power in the relationship of stress (τ) and shearing rate ($\dot{\gamma}$) for a given polymer: $\tau = K\dot{\gamma}^n$. The semiempiric analysis gives the following expression for the speed of radial migration of filler particles:

$$U_r = \frac{1}{50}(n-1)\frac{V}{R^2}\left(\frac{a}{R}\right)^n\left(\frac{r}{R}\right)^{2(n-1)}$$

As can be seen from this expression, there is no radial migration for Newtonian fluid ($n = 1$); for pseudoplastic fluid ($n < 1$), the migration occurs from the tube centre; for dilatant fluid ($n > 1$), the migration is directed towards the tube centre.

Gauthier et al. [38] studied radial migration of spherical particles with a concentration of 5% by volume in the mixture for the following values of parameters: $a = 0.019$ cm; $R = 0.4$ cm; $V = 0.127$ cm^3/s, $n = 0.64$. At the start of the time period, a particle deficit was observed at the channel walls, three minutes later the particles were uniformly distributed in the channel, ten minutes later 75% of particles

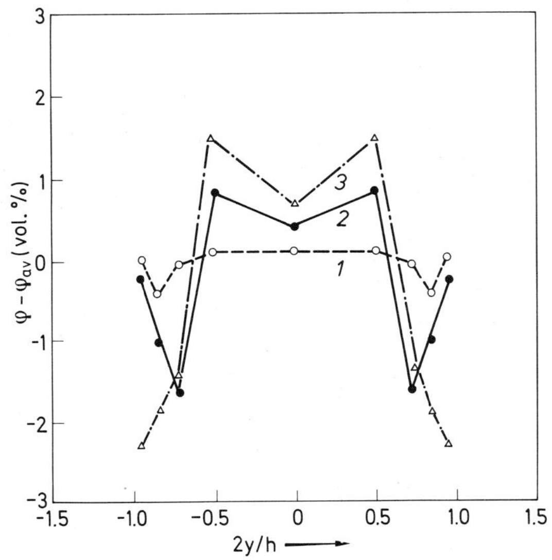

Fig. 4. Distribution of difference between local (φ) and average (φ_{av}) concentrations of spherical filler (with radius a) in cross-section of moulded sample: *1* — a = 0.00011 cm; φ_{av} = 23.6 per cent: *2* — a = 0.0154 cm; φ_{av} = 6.5 per cent; *3* — a = 0.0154 cm; φ_{av} = 24.4 per cent [42]

were in the region $r/R > 0.5$, 30 minutes later about 70% of particles were in the boundary zone $0.7 < r/R < 1$.

This example shows how the distribution of filler particles can vary in the channel cross-section during injection moulding.

Figure 4 shows the results of similar experiment conducted by Hegler and Mennig [42] in a study of the distribution of glass spheres during the course of injection moulding of their mixture with polymer Luran 378P (copolymer of styrene and acrylonitrile) in samples of dumbbell-like shape. The Figure also provides for relationships of deviation in the local concentration of glass spheres from the average (φ_{av}) on the coordinate along sample cross-section (sample thickness h = 0.33 cm). As can be seen, distinct areas of particle concentration are observed for larger spheres (curves 2, 3) whereas such areas are lacking for small spheres (curve 1) and the form of the relationship assumes a lack of filler redistribution.

The Emerman quantitative model [40] makes it possible, with sufficient accuracy, to predict the redistribution of filler particles depending on the parameters of processing and to explain the existing (still not extensive) experimental data. Good agreement should be noted between the experimental data and those calculated according to the Emerman model [40].

4.2 Profile of Conductivity in the Process of Injection Moulding

The results of the above section show that the significant nonuniformity of the distribution of the filler particles in the thickness of sample is observed during injection moulding of the filled polymers. This nonuniformity must affect the electrical properties of CCM owing to the strong dependence of the CCM conductivity on the filler concentration. Although there are no direct comparisons of the concentration profiles and conductivity in the publications, there is data on the distribution of conductivity over the cross-section of the moulded samples.

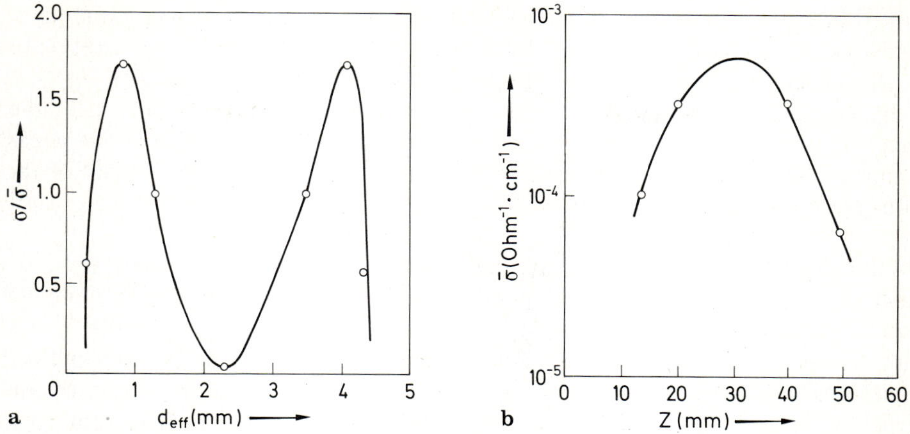

Fig. 5a, b. Conductivity profile in cross-section (*a*) and along (*b*) the sample prepared by injection moulding: $\varphi = 4$ per cent; $\bar{\sigma} = 4 \times 10^{-4}$ Ohm cm^{-1}; a: z = 25 mm [43]

The authors of Ref. [43] studied CCM based on linear polyethylene (M_w — 4.5×10^5) and carbon black XE2, with the filler concentration from 1 to 7.5 per cent by volume. Samples of two types were compared: samples manufactured by pressing and by the injection moulding method with orientation in flow (Elongational Flow Injection Molding [44]). Figures 5 and 6 give the dependences of the averaged cross-section conductivity on the length Z along the flow direction as well as the conductivity profiles for Z = 25 mm for two concentrations of the filler — 4 and 7.5 per cent by volume. As is seen, with lower carbon black concentration, a zone of higher conductivity is observed in the sample in the form of a ring. An elevated birefringence, which is indicative of a highly oriented polymer due to shearing deformations, is observed in the same zone. The conducting carbon black chains also become oriented in their

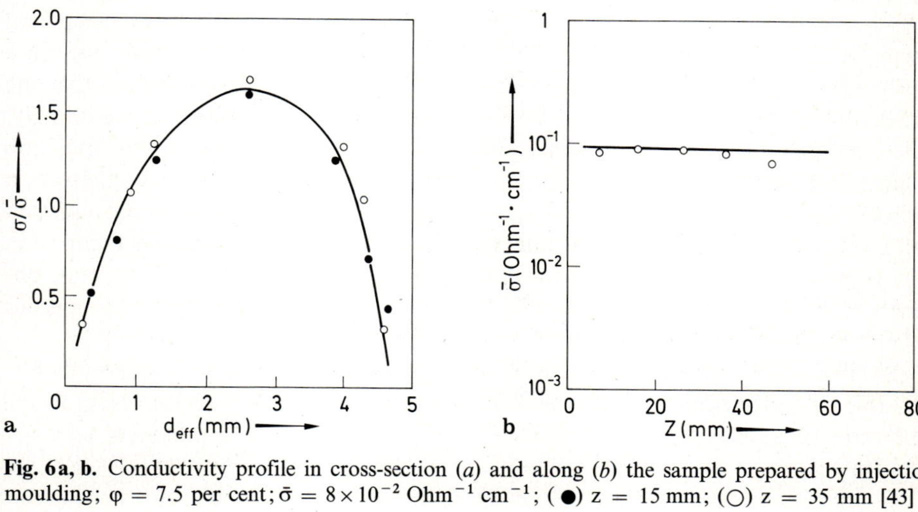

Fig. 6a, b. Conductivity profile in cross-section (*a*) and along (*b*) the sample prepared by injection moulding; $\varphi = 7.5$ per cent; $\bar{\sigma} = 8 \times 10^{-2}$ Ohm^{-1} cm^{-1}; (●) z = 15 mm; (○) z = 35 mm [43]

zone which means higher conductivity. As far as the distance from the injection point increases, the degree of orientation drops down, the zones converge, degenerating to a cylinder, and the averaged cross-section conductivity decreases (Fig. 5b).

Such oriented composites should also have electrical anisotropy. Indeed, for the composite material with $\varphi = 4\%$ of volume $\sigma_{||} = 8 \times 10^{-4}$ Ohm^{-1} cm^{-1}, whereas $\sigma_{\perp}$ (perpendicular to axis Z) is smaller by a factor of 10^3. The conductivity of the isotropic (pressed) sample amounts to 5×10^{-5} Ohm^{-1} cm^{-1} which is also below $\sigma_{||}$.

With the higher carbon black concentration (7.5 per cent by volume), the conductivity profile has a wide maximum and σ does not depend on the length along the flow direction (Fig. 6). This is explained by the fact that the mixture viscosity grows with the filler concentration, as a result of which the shearing zone expands and occupies a large portion of the sample cross-section, and the conductivity in this case reaches 0.1 Ohm^{-1} cm^{-1}. The anisotropy coefficient drops with carbon black concentration growth and for $\varphi = 7.5\%$ by volume the value of $\sigma_{\perp}$ is less than $\sigma_{||}$ by an order of magnitude.

The Emerman model described in the previous section is hardly applicable to the carbon black-filled CCM as the black particles have sizes of hundreds ångström and such a composite, compared with the molding channel size, may be considered as a homogeneous viscous fluid. Therefore, the polymer structure, crystallinity and orientation play an important role for such small particles. The above-given example of manufacture of the CCM demonstrates the importance of these factors being considered during processing of a composite material to and article with the desired electrical properties.

4.3 Distribution of Filler in Conducting Polymer Composites Based on Polymer Blends

As was noted earlier, with the uniform distribution of the conducting filler in a matrix, it is necessary to introduce large amounts of the filler so as to reach the desired level of composite conductivity, which entails aggravation of mixture processing and mechanical properties of the conducting polymer composite. To reduce the filler concentration, it should be segregated, localized in the limited volumes so that the filler concentration in these regions becomes higher than the average one, accordingly, the probability of forming of the conducting channels and conductivity itself are higher in these regions, and this entails the rise of the composite conductivity as a whole. To achieve this goal different methods can be used, for instance, introduction of a neutral filler such as kaolin, talc, etc. [3]. Such fillers decrease, in a composite, the volume of a region accessible for the conducting filler. Nevertheless, general concentration of the fillers remains high and this does not improve mechanical properties of the conducting polymer composite.

But there is another method — the use of heterogeneous blends of polymers [45, 46]. To this end, electrical properties and distribution of the filler (carbon black) in the mixtures of polyethylene and thermodynamically incompatible polymers were investigated.

Compositions were prepared by mixing in the molten state on microrolls using four methods: (1) — the filler is preliminarily introduced into polyethylene and then is mixed with another polymer; (2) — the filler is preliminary introduced into a polymer and then is mixed with polyethylene; (3) — the filler is introduced into the polymer mixture; (4) — a part of the filler is introduced into polyethylene, another part into the polymer in the proportion equal to the contents of the polymers in the mixture, and then all the components are mixed together. Polypropylene, copolymer of ethylene with vinyl acetate and others were used as polymers that are thermodynamically incompatible with polyethylene.

The study of filler distribution by the methods of optical and electronic microscopy has shown that in all compositions obtained by method 4 the filler is distributed rather uniformly as in an individual polymer. In the mixtures of incompatible polymers, obtained by methods 1 and 2, the filler is distributed nonuniformly and there are zones of high concentration of the filler and almost empty ones. The size of such zones is close to the size of polymer regions known for mixtures of thermodynamically incompatible polymers — 1 to 10 μ.

In the case of the filler localization in one of the polymer components of the mixture, an increase of the portion of the second unfilled polymer component in it entails sharp (by a factor of 10^{10}) rise of σ in the conducting polymer composite. In this case the filled phase should be continuous, i.e. its concentration should be higher than the percolation threshold.

One more fact, important in practice, lies in that σ of the compositions based on heterogeneous blends of polymers obtained by the method 3, depends considerably on mixing temperature T_m. This is bound up with a variation of the polymer viscosity with the temperature: on being introduced into the polymer mixture, a filler becomes distributed mostly in the less viscous polymer and, if the viscosity of polymers is almost the same, it is distributed comparatively uniformly and σ of the composition decreases. Therefore, the dependence of σ of the conducting polymer composite on T_m has a minimum (by a factor of 10^2 to 10^4) in the T_m region when the viscosities of the polymer components are close.

It is well known, that under industrial conditions a method of introducing filler into the polymer mixture is used. In this case, the filler is introduced in the form of paste containing up to 60 per cent water in order to reduce viscosity. As heating is affected by viscous friction, the temperature conditions are not stable on mixing and, therefore, conductivity of the conducting polymer composite becomes unreproducible. Up to now this factor has not been taken into consideration.

It should be noted that on mixing by method 1, σ of the composition does not depend on T_m and, thus, is more stable in relation to the variations of mixing process parameters.

The above-described laws of filler distribution in heterogeneous mixtures of polymers are true when the particle size is significantly less than the size of the polymer zones in such mixtures (1 to 10 μ). So, powders of graphite and molibdenum ($S_s =$ $= 2\ m^2/g$) are distributed equally uniformly in all the studied mixtures of polymers irrespective of the mixing conditions for in this case the particle size is comparable with the size of the polymer zones.

Thus, the use of heterogeneous blends of polymers is a successful example of creating the ordered structure of the filler distribution: conductance occurs when the filler concentration exceeds the threshold φ_f in the polymer phase the concentration of which in its turn is higher than the threshold φ_p. In other words, the percolation threshold in such a system equals $\varphi_f \cdot \varphi_p$ which is considerably less than the threshold φ_f for the case of the homogeneous matrix.

The industrial techniques of processing of filled polymer compositions are based on the effect of significant shearing deformations that destroy the current-conducting chains of filler particles. With an increase of shear speed $\dot{\gamma}$, σ of the composition drops to a definite limit (σ_{min}) minimal for a given composition, which is reached for $\dot{\gamma} = 10$ to $30 \, s^{-1}$ [45, 46] for the most conducting polymer composites. For some types of carbon fillers, carbon black in particular, a chain (secondary) structure has thixotropic properties and, when the action of shearing deformation stops, it gradually becomes restored by the action of van der Waals forces, the lower the polymer viscosity, the greater the speed of restoration.

However, the relaxation process is rather slow whereas thermoplast processing calls for almost instantaneous cooling of the melt after molding and due to this fact the chain structure is not restored. From this it follows that in molding a conducting polymer composite the cooling of the melt should have a speed comparable to the speed of restoration of the filler secondary structure, otherwise the conductivity of the polymer composite will be reduced.

4.4 Some Features of Processing of Composites with Conducting Fibers

As already noted, the main merit of fibers used as a filler for conducting composite materials is that only low threshold concentrations are necessary to reach the desired level of composite conductivity. However, introduction of fiber fillers into a polymer with the help of ordinary plastic materials processing equipment presents certain difficulties which are bound up mainly with significant shearing deformations entailing fiber destruction and, thereby, a decrease of parameter l/d which determines the value of the percolation threshold.

Destruction of the fiber filler may occur during compression of a composite. It is evident that the destroying process is more pronounced for high-modulus, brittle fibers such as glass or carbon. For the materials containing organic fibers the destruction rate is considerably lower.

The analysis of the distribution curves of the fiber filler length after compression permits one to conclude that a variation of the fiber average length at compression may be approximately considered as a function of the value of applied pressure irrespective of the composition of the mixture and the state of the polymer [47]. In this case, it should be taken into consideration that longer fibers are destroyed more easily. This is bound up with destruction due to bending at the fiber contact points, the number of which depends directly on the fiber length.

Destruction of fibers under the conditions of shearing deformations of the composition melt [48] has a somewhat different nature — such as those conditions arising in up-to-date, plastic materials processing installations. In this case two characteristic regions may be distinguished in the composition melt: a boundary layer with high shearing speeds and reduced filler concentration, and the basic mass of the melt outside this layer in which the filler concentration and, thereby, viscosity are higher than in the boundary layer. This prevents the development of significant shearing deformations. In such a model, filler fibers may be considered as long flexible bars secured in the composition mass and projecting into the moving mass of the boundary layer. In this case two forces are acting on the fiber, namely, drag F_d and force F_τ caused by the tangential stresses acting on the sections of fibers parallel to the direction of flow (the analog of such a model is exemplified by the bending of a thin tree stem in a strong wind).

The nature and place of destruction are different for low- and high-modulus fibers. The most probable kind of destruction of low-modulus fibers is their destruction during stretching in the section parallel to the direction of flow. For high-modulus fibers this kind of destruction is improbable because of their high strength. Apparently, the most probable destruction for such fibers may occur due to bending stresses in places of their "fit" at the boundary layer caused by bending moment M_o arising from forces F_d and F_τ.

Such a model makes it possible to calculate a change of fibers distribution along the length in the boundary layer. At present, practically the sole approach to the analysis of destruction when the fiber filler flows in the basic mass, outside the boundary layer, is an experimental determination of destruction kinetics for a given pair — fiber filler and polymer. Such dependencies can be obtained with the help of, say, rotary viscosimeters [47].

This makes it possible to conclude that the intensity of fiber filler destruction depends on the zone of processing installation where the filler is introduced (Fig. 7). As the Figure shows, with glass fibers introduced into the loading zone, their destruction occurs rapidly (curve 1). When fibers are introduced into the melt zone of an extruder the fiber destruction is less intensive.

The influence of processing conditions on material electrical properties can be traced in the following example [16]. From a composition containing 6 per cent by weight of stainless-steel fibers (diameter 7 μ) two plates were manufactured by the injection moulding method under different processing conditions. In the first sample the average value of 1/d amounted to 200, the value of 1/d being equal to or more than 200 for more than 60 per cent of the fibers. In the second sample, only 22 per cent of the fibers had an aspect ratio of more than 200 and the average value of 1/d amounted

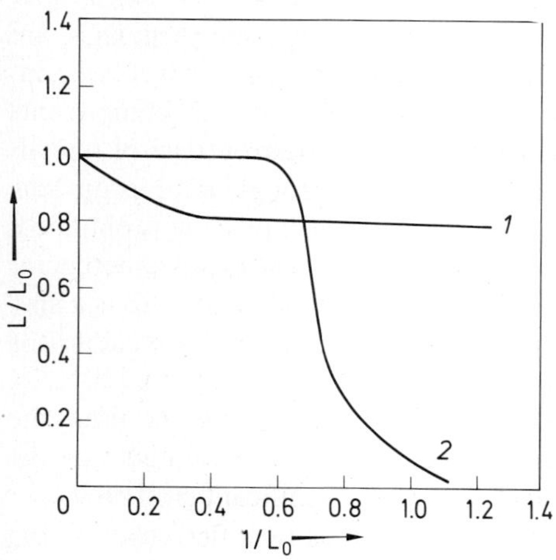

Fig. 7. Relationship of average fiber length $\bar{L}/L_0$ and screw length $1/L_0$ of extruder: *1* — introducing fibers into melt zone; *2* — introducing fibers into leading zone [47]

to 120. In this case the conductivity of the first plate was 15 times higher than that of the second.

Special measures are taken to facilitate processing and decrease the probability of steel fiber destruction. Fibers are used in the form of blanks, bundle lengths 3 to 8 mm long, consisting of 1159 individual fibers in a polymer binder the portion of which can amount to 6 to 30 per cent of weight. Since fibers have a regular cylindrical shape and, thereby, do not form agglomerates the main problem is fiber destruction. In order to avoid destruction, fiber blanks are introduced into the melt zone of an extruder and a polymer which is compatible with the composite polymer matrix in terms of its chemical and rheological properties and melting temperature, is chosen as a binder.

As a result, the merits of fiber fillers for the manufacture of conducting composite materials by extrusion and injection moulding methods can be realized sufficiently.

4.5 Polymerization Filling — A Novel Technique for Manufacturing Conducting Composite Materials

Quite naturally, novel techniques for manufacturing composite materials are in principal rare. The polymerization filling worked out at the Chemical Physics Institute of the USSR Academy of Sciences is an example of such techniques [49–51]. The essence of the technique lies in that monomer polymerization takes place directly on the filler surface, i.e. a composite material is formed in the polymer forming stage which excludes the necessity of mixing constituents of a composite material. Practically, any material may be used as a filler; the use of conducting fillers makes it possible to obtain a composite material having electrical conductance. The material thus obtained in the form of a powder can be processed by traditional methods, with polymers of many types (polyolefins, polyvinyl chloride, elastomers, etc.) used as a matrix.

Natural graphite and synthetic graphite were used as fillers for the manufacture of conducting composite materials by the polymerization filling technique [24, 53–56]. The manufacture of conducting polymer composite materials by this technique on the basis of some kinds of carbon black is also known [51, 52].

To obtain a graphite- and polypropylene-based composite material the Ziegler and Natta catalysts are used [57]. To this end, first $(C_2H_5)_2AlCl$ and then $TiCl_4$ are applied from the gas phase onto the graphite preevacuated at about 200 °C (the investigations have shown that another sequence of application of catalyst components entails the decrease of composite material conductivity by several orders of magnitude). Therefore, activated filler is introduced into liquid propylene or propylene solution where polymerization occurs. As a result, a powder-like product is obtained, with the polypropylene applied onto the graphite being noted for high stereoregular properties and having an atactic fraction of 3 to 4 per cent of the mass. This is a specific feature of graphite as support material on which the polymerization catalyst is secured [58].

The degree of filling depends on polymerization time: the greater the time, the thicker the layer of polymer on the surface of the filler particles, and the smaller the filling degree which, thus, can be regulated in a wide range from less than 1 to 90 per cent by weight. And the composite material obtained by the polymerization filling technique and containing up to 85 per cent by weight of graphite retains admissible mechanical properties, while the composite material containing the same amount of graphite but obtained by ordinary mixing methods (blade mixer and laboratory extruder) has no mechanical stability, i.e. falls apart.

The comparison of conductivity of polypropylene- and natural graphite-based composite materials obtained by polymerization filling and mechanical mixing (See Fig. 2) has shown that, given an equal concentration of graphite, the latter have a much lower conductivity (samples for conductivity measurements were prepared by pressing). With a graphite concentration of about 10 per cent by volume the difference reaches a factor of 10^8 and this difference drops with filler concentration increase to reach a factor of 10^2 at a concentration of about 50 per cent by volume. It should be noted, that at a given concentration σ of the composite material obtained by polymerization filling amounts to about 10^2 Ohm^{-1} cm^{-1} which can not be achieved if the traditional methods of mixing carbon fillers are used.

The important parameter, both from the viewpoint of conductance mechanism and practical utilization of composite materials is the temperature dependence of σ which may be characterized by the energy of activation of conductivity or temperature coefficient of resistance (TCR). It turned out that, for composite materials obtained by polymerization filling, the value of TCR in the temperature range from 4.2 to 300 K is not higher than 1×10^{-4} K^{-1} (activation energy is about 0.5 eV), which is considerably lower than for compositions obtained by mechanical mixing or for the known conducting composities [59]. In this case the material withstands multiple thermal cycles from liquid helium temperature to room temperature and vice versa, which is indicative of high mechanical stability of the composite. Pronounced decrease of σ occurs only at temperatures higher than 353 K but even in this region the smaller the concentration of graphite in a composite material the larger is the value of TCR.

Among other properties, distinguishing natural graphite-based CPCM obtained by polymerization filling from composite materials obtained by mechanical mixing, the following should be noted: (1) a linearity of current-voltage characteristic up to current density of about 1 A/cm^2; (2) a high degree of homogeneity of electrical properties within the material volume; (3) higher thermal conductivity (up to 1 W/(m.K)) with graphite concentration being 30 per cent of mass; (4) high physical-mechanical properties; (5) unusual magnetic properties [78, 79].

The nature of a change of mechanical properties is somewhat unusual for polymerization-filled composite materials with the increase of natural graphite concentration [56]. The modulus of elasticity of the composite material increases fivefold with graphite concentration C changing from 0 to 60 per cent of mass (for mechanically obtained mixtures the value of modulus is 1.6 times lower). In this case, elongation at rupture drops naturally, reaching 5 per cent for $C = 10\%$, and at an increase of C the material becomes short-brittle. The tensile strength and compressive yield point are practically constant up to graphite concentration of 70 per cent and the polymerized composite materials do not collapse even up to $C = 70\%$ (a sample becomes barrel-shaped) whereas mechanically-obtained mixtures are short-brittle even at $C = 10\%$ of mass. The unusual physical-mechanical properties of composite materials obtained by polymerization filling may be obviously explained by stronger adhesive interaction of the filler and polymer matrix owing to the essence of the technique — chemical grafting of polymer to a filler.

The most probable reason for improvement of the properties of composite materials obtained by the polymerization filing, as compared with mechanical mixtures, lies in the peculiarities of a structure of natural graphite and the polymerization process. It is known that natural graphite represents sufficiently large crystals of a laminated structure, the individual lamellas are weakly interlinked owing to a large distance between them [60]. During polymerization a catalyst can penetrate between the separate lamellas, and the forming polymer pushes away the lamellas thus weakening their bonds to a greater extent. So during polymerization filling a splitting of natural graphite particles to separate flakes, that have large aspect ratio, occurs. The data from X-ray examination [61] and measurements of anisotropy coefficients of the electrical and thermal conductivity [24] of these materials, that reach 20 (whereas for mechanical mixtures its value is not in excess of 4) point to this fact.

The data obtained during the use of graphite of another structure confirm the proposed explanation. Thus, the properties of a composite material do not depend on the manufacture technique [54] if in the capacity of a filler a synthetic graphite was used, the particles of which consisted of strongly bound microcrystallites about 200 Å in size, incapable of flaking. The use of the natural graphite the particles of which were ground to the average size of about 200 Å [61] leads to the same conclusion.

Alternative explanations of the high conductivity of composite materials obtained by polymerization filling are given in works [62, 63] where conductivity higher than that of the graphite proper is attributed to a polymer interlayer between graphite particles, are, in our opinion, insufficiently convincing and cannot explain the whole of the experimental data.

5 Applications of Conducting Polymer Composite Materials

Any review devoted to conducting composites would be incomplete if the application fields of such composites were not described even if briefly. One of the first, if not the foremost, examples of the utilization of the CPCM is antistatic materials [1]. For the materials of this kind resistivity ϱ of less than 10^6 to 10^8 Ohm cm is not required, and this is achieved by introducing small amounts (several per cent) of a conducting filler, say, carbon black [4].

The CPCM are widely used as low-temperature heaters [3]. Having sufficiently high resistivity compared to metals, such heaters possess a small specific power and their fire hazard is lower. At present, the task of designing a heater with temperature self-limitation is attracting great attention. In such heaters the power output sharply drops upon reaching a preset temperature and again rises after cooling. Heaters with these characteristics do not need regulating equipment, therefore, they are relatively inexpensive and convenient in service. They can be used, for example, for heating vehicle storage batteries or oil pipelines operating at low temperatures. The material for such heaters should have a temperature coefficient of resistance (TCR) which drastically rises in the required temperature range, and this calls for the use of the composites the properties of which, including the TCR, can be regulated within the wide ranges by selecting a matrix, filler, mixing method and other parameters [3].

The use of corona-forming and precipitation electrodes made of a composite based on polypropylene and acetylene black (8 per cent by volume) is among the examples of nontraditional utilization of CPCM [64–66]. For extrusion mixing the ϱ_v of composite material equals 3.8×10^2 Ohm cm. Electrode parts manufactured of the composite materials have undergone laboratory and industrial tests and shown good operability in long service. The effect of using such a material in electrical filters is rather high as it substitutes large amounts of nonferrous metals, lead in particular, that are usually used for the purpose.

One of the most promising, and, apparently, large-scale, applications is shielding of electromagnetic radiation [14–16, 67, 68]. Shielding, in a broad sense, is protection of electronic devices and circuits against interferences in the form of external electromagnetic fields of different frequency [69, 70]. At present, this problem is becoming more pressing owing to the growing concentration of electronic devices, computers, the increase of wiring density, and intricacy of electronic circuits, i.e. the increase of mutual influence of various devices, the casings of which are made mostly of plastic materials, which leads to failures filled with serious consequences. The acuity of the problem compelled the USA Federal Communication Commission to legislatively limit the electromagnetic noise level created by electronic devices in 1979. Beginning with October 1, 1983 all such devices must meet the requirements of FCC standard No. 20780 [67]. According to this standard, all electronic devices are divided into classes: class A — devices used in industry, laboratories, etc.; class B — domestic devices. The limitations cover the magnitude of intensity of electrical field of different frequency, created by a device at a certain distance. The limitations for class A devices are stricter as their concentration under the conditions they are used in is considerably higher than, say, in an apartment flat.

The idea of using CPCM for shielding is rather alluring. Indeed, a casing of an article or instrument manufactured of such a material serves at the same time as a screen to protect against electromagnetic radiation. All the above-described operations involved in applying additional layers become unnecessary.

Therefore, in connection with the introduced limitations of electromagnetic noise level, in the early 1980 s, a vast field of application seemed open for conducting composites. In the USA more than 80000 t of various plastic materials is consumed for the manufacture of casings of commercial electronic devices and almost all the casings call for screening [71]. But in reality these hopes were not realized. Table 1 shows the percentage of different ways of manufacture of screens in 1982 and 1986.

As shown, drastic changes in the ratios of different shielding techniques occurred over the four years and at present the majority of screens are manufactured by applying conducting coatings (paints) while the share of conducting composites is very small: in 1986 it amounted to 500 t in absolute magnitude [71].

Table 1. Percentage Use of various methods of Manufacturing Electromagnetic Screens

Method	Percentage	
	1982	1986
1. Conducting coatings (paints)	40	75
2. Arc-discharge spraying	49	4
3. Vacuum deposition	4	2
4. Electrochemical coating	—	8
5. Conducting plastic materials	—	0.5
6. Miscellaneous	7	10.5

The main reasons for this lie in feasibility. Conducting fillers are rather expensive and their use increases the cost of an article. Besides, filled polymers have worse physical-mechanical properties, especially impact strength and flexural modulus. The use of fillers is also detrimental to the articles appearance and calls for additional treatment. The continuous development of electronics has also contributed to a loss of interest to conducting composites as screening materials: the improvement of components and circuits of devices made it possible to reduce currents consumed and, thereby, noise level; a so called "can" method is practised on a wide scale in order to cover the most sensitive or noisy sections of a circuit with metal housings [14].

Shielding capability of a composite material depends mainly on its conductivity.

Table 2 provides different fillers and approximate concentrations at which composites have surface conductance of about 1 Ohm^{-1} which corresponds to a shielding level of about 40 dB [14].

Table 2. Types of Conducting Fillers and Concentrations Corresponding to Shielding Level of 40 dB and Surface Conductivity of 1 Ohm^{-1}

Filler	Concentration in composite, %
Aluminium	
fibers	30
flakes	40
powder	30
glass fiber coated with Al	30
Carbon	
commercial carbon	35
commercial furnace carbon black	30 to 40
carbon black Ketjenblack EC	5 to 15
Polyacrylonitrile-based carbon fibers	30
Copper powder	35
Metallized glass fibers	40
Carbon fibers coated with Ni	10
Steel fibers	5

Thus, bearing in mind that smaller filler concentrations worsen the physical-mechanical properties of the composites to a smaller degree, it follows that the most promising are the fillers which provide for a preset level of σ at smaller concentration. The table shows that the most promising are carbon fibers coated with Ni (American Cyanamid) and steel fibers (Brunswick Corp.) [16].

The main problem in using composites with conducting fibers lies, as was noted, in the necessity of precluding the destruction of fibers during mixing because, at a given concentration, the larger the ratio of fiber length and fiber diameter (aspect ratio) the higher the σ of the composition [11, 16]. Strong steel fibers (7 μm in diameter) resist destruction better than brittle carbon fibers and, therefore, they are considered as promising candidates for the manufacture of composites for electromagnetic screens [8, 16].

The influence of aspect ratio of steel fibres on effective shielding may be illustrated by the following example from Sect. 4.4. A composite with an average aspect ratio of fibres of 200 has σ 15-fold higher (with the same concentration) than that of a composite with average aspect ratio of 120. Accordingly, the effective shielding amounts to 37 and 17 dB [16].

Particular examples of using polymer composites as screens are given in [14–16, 67–75]. The present review does not touch the properties of the composite materials based on fabrics of conducting fibres due to the fact that manufacturing techniques for such materials are specific and differ greatly from the mixing processes considered above. However, these materials also have an application field, say, in contacts for calculator and computer keyboards [9] and even in small-power electric motor commutators as a partial substitute for copper [76, 77].

6 Summary

Despite the existing opinion that the production of the conducting polymer composites is close to art or even to black magic [14], we have tried to show that production is based on sufficiently clear physical principes and, therefore, the material properties can be calculated and controlled. The complexity of the task lies in the necessity of allowing for many factors, but having overcome the difficulties we will have an opportunity of creating materials possessing novel properties or capable of substitutig for materials the resources of which are nonrenewable.

7 References

1. Norman RH (1970) Conductive rubbers and plastics, Appl. Sci. Pub., Essex, England
2. Sichel EK (1982) Carbon black — Polymer composites, Marcel Dekker, New York
3. Gul' VE, Shenfil LZ (1984) Conducting polymer composites, Moscow, Chemistry (in Russian)
4. Vasilenok YN (1981) Prevention of statistical charging of polymers, Leningrad, Chemistry (in Russian)
5. Gurland JC, Tanner DB (eds) (1978) Electrical Transport and Optical Properties of Inhomogeneous Media AIP, New York
6. Seymour RB (1984) Conductive polymers, Plenum, New York
7. Conducting Polymer Symp., Las Vegas, NV, Aug. 26–27, 1980, Plenum New York, 1981
8. Crossman RA (1985) Polym. Eng. Sci. 25: 507
9. Mair HJ, Roth S (eds) (1986) Elektrisch leitende Kunststoffe, Carl Hanser, Munchen
10. Shevchenko VG, Ponomarenko AT (1983) Advances in Chemistry, USSR, 52: 1336
11. Shevchenko VG, Ponomarenko AT, Enikolopyan NS (1985) Plaste und Kautschuk 32: 401
12. Ponomarenko AT, Shevchenko VG, Tchmutin IA, Ovchinnikov AA, Enikolopyan NS (1986) Conducting polymer composites, In: Sedlacek B (ed) Polymer composites, Walter de Gruyter, B., N.Y., 1986, p. 67.
13. Katz HS, Milewski JV (eds) (1978) Handbook of fillers and reinforcements for plastics, Van Nostrand Reinh., New York
14. Kiesche E: Plast. Techn., Nov 1985, p 77
15. Bigg DM, Stutz DE (1983) Polymer Comp. 4: 40
16. Tolokan RP, Nablo JC: Plast. Eng., Aug 1985, p 31
17. Levit RM (1986) Conducting Chemical Fibers, Moscow, Chemistry (in Russian)
18. Sheng P (1980) Phys. Rev. 21B: 2180
19. Malliaris A, Turner DT (1971) Journ. Appl. Phys. 42: 614
20. Broadbent SR, Hammersley JM (1957) Proc. Cambr. Soc. (1957) 53: 629

21. Shante VKS, Kirkpatrick S (1971) Adv. Phys. 20: 325
22. Kirkpatrick S (1973) Rev. Mod. Phys. 45: 574
23. Shklovsky BI, Efros AL (1975) Advances in Physical Sciences, USSR, 117: 401
24. Galashina NM, Shevchenko VG, Ponomarenko AT, Grigorov LN, Shklyarova EI, Djachkovsky FS, Enikolopyan NS (1985) In: Ovchinnikov AA (ed) Electronics of organic materials Moscow Science, p 43 (in Russian)
25. Sazhin BI (ed) (1970) Electrical properties of polymers, Leningrad Chemistry (in Russian)
26. Sheer GE, Turner DT (1971) In: Gould RF (ed) Multicomponent polymer systems, ACS, Wash.
27. Yun-ichi Y, Osamu M, Yuzo K (1978) Rev Elec. Commun Lab. 26: 616
28. USA patent No. 382942 (1983)
29. Tikhomirov AF, Pugachev AK, Olshevsky OI, Sazhin BI (1988) Plastic Materials, USSR No. 5, p 13
30. Bhattacharya R (1982) Polym.-Plast. Technol. Eng. 19: 21
31. Davenport DE (1981) In: Conduct. Polym. Proc. Symp., Las Vegas, 1980, Plenum, New York, p 39
32. Bazarova FF, Kolosova NS (1975) Cements in production of radioelectronic equipment, Moscow, Energy (in Russian)
33. Klason C, Kubat J (1975) Journ. Appl. Polym Sci. 24: 831
34. Fox L (1978) RCA Review 39: 119
35. Skutnik Bolesh J (1984) Rubber Chem. Technol. 57: 403
36. Torner RV (1972) Basic processes of polymer processing (Theory and calculations methods), M., Chemistry (in Russian)
37. Gauthier F, Goldsmith HL, Mason SG (1971) Rheol. Acta 10: 344
38. Gauthier F, Goldsmith HL, Mason SG (1971) Trans. Soc. Rheol., 15: 297
39. Brunn P (1980) Journ. Non-Newt. Fluid Mech. 7: 271
40. Emerman SH (1987) Polym. Eng. Sci. 27: 1105
41. Kubat J, Szalanczi A (1974) Polym. Eng. Sci. 14: 873
42. Hegler RP, Mennig G (1985) Polym. Eng. Sci. 25: 395
43. Bayer RK et al. (1988) Journ. Mater. Sci. 23: 475
44. Bayer RK, Eliah AE, Seferis JC (1984) Polym. Eng. Rev. 4: 201
45. Pavlii VG, Zaikin AE, Kuznetsov EV (1984) Proceedings of XII All-Union Conference on Organic Semiconductors, Moscow, Patent publishers, pp 117–118 (in Russian)
46. Pavlii VG, Zaikin AE, Kuznetsov EV (1986) Higher School Bulletin, USSR, Chemistry and Chemical Technology, 29: 84
47. Kim VS, Skachkov VV (1980) Dispersion and mixing in the processes of manufacture and processing of plastic materials, Moscow, Chemistry, p 240 (in Russian)
48. Bachellor GK (1969) Introduction to fluid dynamics, Wiley N.Y.
49. Enikolopov NS (1980) Nature, USSR 8: 62
50. Dyachkovsky FS, Novokshonova LA (1984) Advances in Chemistry, USSR, 53: 200
51. Baulin AA, Krasnoshchekov AM, Deyanova AS, Vasilenok YI (1982) Applied Chemistry Journal, USSR 55: 2534
52. Aivazyan FN, Matkovsky PE, Ponomarenko AT, Pavlii VG, Shamsullin IF, Kolesnikov YN, Enikolopyan NS (1983) In: Complex metalloorganic catalysts of elefin polymerization, Chernogolovka, p 73 (in Russian)
53. Enikolopyan NS, Galashina NM, Shevchenko VG, Ponomarenko AT et al. (1984) USSR Certificate of Authorship No. 1240761
54. Nedorezova PM, Shevchenko VG, Grinev VG, Galashina NM, Tsvetkova VN, Ponomarenko AT, Dyachkovsky FS, Levintovich PY, Katasonov AS, Enikolopyan NS (1985) Collection of Reports at II All-Union Conference "Highly Filled Complex Polymer Materials in the Economy", I., vol 1, pp 20–25 (in Russian)
55. Levintovich PY, Shevchenko VG, Kotosonov AS, Nedorezova PM, Ponomarenko AT (1984) Proceedings of XIII All-Union Conference on Organic Semiconductors, Moscow, Patent, p 96 (in Russian)
56. Galashina NM, Shevchenko VG, Ponomarenko AT, Nedorezova PM, Filippov PG, Grigorov LN, Dyachkovsky FS, Tsvetkova VN, Enikolopyan NS: Abstracts of IUPAC Macro-83, Bucharest, Romania, Spt. 5–9, 1983, Sect. 6–7, p 121

57. Chirkov IM, Matkovsky PE, Dyachkovsky FS (1976) Polymerization with Complex Metalloorganic Catalysts, M., Chemistry, p 414 (in Russian)
58. Galashina NM, Nedorezova PM, Tsvetkova VN, Dyachkovsky FS, Enikolopyan NS (1984) USSR Academy of Sciences Proceedings, 278: 620
59. Sazhin BI (ed) (1977) Electrical properties of polymers, Moscow, Chemistry (in Russian)
60. Ubbelode AR, Lewis FA (1960) Graphite and its crystal compounds, Oxford Univ. Press
61. Nedorezova PM, Galashina NM, Tsvetkova VN, Grinev VG, Shevchenko VG, Ponomarenko AT, Tovmasyan YM, Kolbanov IV, Dyachkovsky FS, Enikolopyan NS: Abstracts of IUPAC Macro-87, Merseburg, June 30–July 4, 1987, p 102
62. Grigorov LN, Shklyarova, Dorofeeva TV, Galashina NM, Enikolopyan NS (1984) USSR Academy of Sciences Proceedings 278: 911
63. Grigorov LN, Galashina NM, Enikolopyan NS (1984) USSR Academy of Sciencies Proceedings, 274: 840
64. Moshkin AA, Losoto AP, Berezhnoy AI, Krasninov AS (1979) Nonferrous Metallurgy, USSR, 21: 34
65. Losoto AP, Moshkin AA, Akutin MS, Berezhnoy AI, Udalova VI, Petrova VF (1981) Plastic Materials, USSR, 3: 29
66. Losoto AP, Akutin MS, Petrova VF, Moshkin AA, Udalova VI, Pegasov VV, Konysheva AV (1978) Nonferrous Metallurgy, USSR, 17: 52
67. Bledzki A, Stankowska D (1984) Kunststoffe 74: 89
68. Mobius KH (1988) Kunststoffe 78: 345
69. Regan J (1982) Polym.-Plast. Techn. Eng. 18: 47
70. Vance EF (1980) IEEE Trans. on Electromagn. Comp. EMC-22, 4: 319
71. Toensmeier PA (1987) Modern Plast. Int. 2: 47
72. Stephens KA (1987) Plast. Eng. 2: 33
73. Carpenter R (1986) Appliance Manufact. 8: 20
74. Yoshitake M (1985) Japan Plast. Age 23: 21
75. Inkpen SL, Melcher JR (1985) Polym. Eng. Sci. 25: 289
76. Bekishev RF, Daneker VA (1979) Electrotechnical Industry, USSR, Electrotechnical Materials series 5: 8
77. USSR Certificate of Autorship No. 997153, Inventions Bulletin, 1983, No. 6
78. Pristupa AI, Shevchenko VG, Ponomarenko AT, Chmutin IA, Ovchinnikov AA, Enikolopyan NS (1986) Proceedings of 8th Conf. Polymerphysik-86, Leipzig, Kurzfassung Vorträge und Poster, p 98
79. Pristupa AI, Petrov NK, Shevchenko VG, Ponomarenko AT, Enikolopyan NS (1986) Proceedings of 8th Conf. — Polymerphysik — 86, Leipzig, Kurzfassung Vorträge und Poster, pp 199–200

Editor: N. S. Enikolopyan
Received Mai 30, 1989

Author Index Volumes 1–96

Allegra, G. and *Bassi, I. W.*: Isomorphism in Synthetic Macromolecular Systems. Vol. 6. pp. 549–574
Andrade, J. D., Hlady, V.: Protein Adsorption and Materials Biocompability: A. Tutorial Review and Suggested Hypothesis. Vol. 79, pp. 1–63
Andreis, M. and *Koenig, J. L.*: Application of NMR to Crosslinked Polymer Systems. Vol. 89, pp. 69–160.
Andrews, E. H.: Molecular Fracture in Polymers. Vol. 27, pp. 1–66.
Anufrieva, E. V. and *Gotlib, Yu. Ya.*: Investigation of Polymers in Solution by Polarized Luminescence. Vol. 40, pp. 1–68.
Apicella, A. and *Nicolais, L.*: Effect of Water on the Properties of Epoxy Matrix and Composite. Vol. 72, pp. 69–78.
Apicella, A., Nicolais, L. and *de Cataldis, C.*: Characterization of the Morphological Fine Structure of Commercial Thermosetting Resins Through Hygrothermal Experiments. Vol. 66, pp. 189–208.
Argon, A. S., Cohen, R. E., Cebizlioglu, O. S. and *Schwier, C.*: Crazing in Block Copolymers and Blends. Vol. 52/53, pp. 275–334.
Argon, A. S. and *Cohen, R. E.*: Crazing and Toughness of Block Copolymers and Blends. Vol. 91/92, pp. 301–352.
Aronhime, M. T., Gillham, J. K.: Time-Temperature Transformation (TTT) Cure Diagram of Thermosetting Polymeric Systems. Vol. 78, pp. 81–112.
Arridge, R. C. and *Barham, P. J.*: Polymer Elasticity. Discrete and Continuum Models. Vol. 46, pp. 67–117.
Aseeva, R. M., Zaikov, G. E.: Flammability of Polymeric Materials. Vol. 70, pp. 171–230.
Ayrey, G.: The Use of Isotopes in Polymer Analysis. Vol. 6, pp. 128–148.

Bässler, H.: Photopolymerization of Diacetylenes. Vol. 63, pp. 1–48.
Baldwin, R. L.: Sedimentation of High Polymers. Vol. 1, pp. 451–511.
Bascom, W. D.: The Wettability of Polymer Surfaces and the Spreading of Polymer Liquids. Vol. 85, pp. 89–124.
Balta-Calleja, F. J.: Microhardness Relating to Crystalline Polymers. Vol. 66, pp. 117–148.
Barbé, P. C., Cecchin, G. and *Noristi, L.*: The Catalytic System Ti-Complex/$MgCl_2$. Vol. 81, pp. 1–83.
Barton, J. M.: The Application of Differential Scanning Calorimetry (DSC) to the Study of Epoxy Resins Curing Reactions. Vol. 72, pp. 111–154.
Ballauff, M. and *Wolf, B. A.*: Thermodynamically Induced Shear Degradation. Vol. 84, pp. 1–31.
Basedow, A. M. and *Ebert, K.*: Ultrasonic Degradation of Polymers in Solution. Vol. 22, pp. 83–148.
Batz, H.-G.: Polymeric Drugs. Vol. 23, pp. 25–53.
Baur, H. see Wunderlich, B.: Vol. 87, pp. 1–121.
Bell, J. P. see Schmidt, R. G.: Vol. 75, pp. 33–72.
Bekturov, E. A. and *Bimendina, L. A.*: Interpolymer Complexes. Vol. 41, pp. 99–147.
Berger, L. L. see Kramer, E. J.: Vol. 91/92, pp. 1–68.

Bergsma, F. and *Kruissink, Ch. A.:* Ion-Exchange Membranes. Vol. 2, pp. 307–362.
Berlin, Al. Al., Volfson, S. A., and *Enikolopian, N. S.:* Kinetic of Polymerization Processes. Vol. 38, pp. 89–140.
Berry, G. C. and *Fox, T. G.:* The Viscosity of Polymers and Their Concentrated Solutions. Vol. 5, pp. 261–357
Bevington, J. C.: Isotopic Methods in Polymer Chemistry. Vol. 2, pp. 1–17.
Beylen, M. van, Bywater, S., Smets, G., Szwarc, M., and *Worsfold, D. J.:* Developments in Anionic Polymerization – A Critical Review. Vol. 86, pp. 87–143.
Bhuiyan, A. L.: Some Problems Encountered with Degradation Mechanisms of Addition Polymers. Vol. 47, pp. 1–65.
Biedermann, H. see Osada, Y.: Vol. 95, pp. 57–110.
Billingham, N. C. and *Calvert, P. D.:* Electrically Conducting Polymers – A Polymer Science Viewpoint. Vol. 90, pp. 1–104.
Bird, R. B., Warner, Jr., H. R., and *Evans, D. C.:* Kinetik Theory and Rheology of Dumbbell Suspension with Brownian Motion. Vol. 8, pp. 1–90.
Biswas, M. and *Maity, C.:* Molecular Sieves as Polymerization Catalysts. Vol. 31, pp. 47 bis 88.
Biswas, M., Packirisamy, S.: Synthetic Ion-Exchange Resins. Vol. 70, pp. 71–118.
Block, H.: The Nature and Application of Electrical Phenomena in Polymers. Vol. 33, pp. 93–167.
Bodor, G.: X-ray Line Shape Analysis. A. Means for the Charactrization of Crystalline Polymers. Vol. 67, pp. 165–194.
Böhm, L. L., Chmelíř, M., Löhr, G., Schmitt, B. J. and *Schulz, G. V.:* Zustände und Reaktionen des Carbanions bei der anionischen Polymerisation des Styrols. Vol. 9, pp. 1–45.
Bölke, P. see Hallpap, P.: Vol. 86, pp. 175–236.
Bormashenko, E. Yu.: see Fridman, M. L.: Vol. 93, pp. 81–136.
Boutevin, B.: Telechelic Oligomers by Radical Reactions. Vol. 94, pp. 69–106.
Boué, F.: Transient Relaxation Mechanisms in Elongated Melts and Rubbers Investigated by Small Angle Neutron Scattering. Vol. 82, pp. 47–103.
Bovey, F. A. and *Tiers, G. V. D.:* The High Resolution Nuclear Magnetic Resonance Spectroscopy of Polymers. Vol. 3, pp. 139–195.
Braun, J.-M. and *Guilett, J. E.:* Study of Polymers by Inverse Gas Chromatography. Vol. 21, pp. 107–145.
Breitenbach, J. W., Olaj, O. F. und *Sommer, F.:* Polymerisationsanregung durch Elektrolyse. Vol. 9, pp. 47–227.
Bresler, S. E. and *Kazbekov, E. N.:* Macroradical Reactivity Studied by Electron Spin Resonance. Vol. 3, pp. 688–711.
Brosse, J.-C., Derouet, D., Epaillard, F., Soutif, J.-C., Legeay, G. and *Dušek, K.:* Hydroxyl-Terminated Polymers Obtained by Free Radical Polymerization. Synthesis, Characterization, and Applications. Vol. 81, pp. 167–224.
Bucknall, C. B.: Fracture and Failure of Multiphase Polymers and Polymer Composites. Vol. 27, pp. 121–148.
Burchard, W.: Static and Dynamic Light Scattering from Branched Polymers and Biopolymers. Vol. 48, pp. 1–24.
Bywater, S.: Polymerization Initiated by Lithium and Its Compounds. Vol. 4, pp. 66–110.
Bywater, S.: Preparation and Properties of Star-branched Polymers. Vol. 30, pp. 89–116.
Bywater, S. see Beylen, M. van: Vol. 86, pp. 87–143.

Calvert, P. D. see Billingham, N. C.: Vol. 90, pp. 1–104.
Candau, S., Bastide, J. und *Delsanti, M.:* Structural. Elastic and Dynamic Properties of Swollen Polymer Networks. Vol. 44, pp. 27–72.
Carrick, W. L.: The Mechansim of Olefin Polymerization by Ziegler-Natta Catalysts. Vol. 12, pp. 65–86.
Casale, A. and *Porter, R. S.:* Mechanical Synthesis of Block and Graft Copolymers. Vol. 17, pp. 1–71.

Cecchin, G. see Barbé, P. C.: Vol. 81, pp. 1–83.
Cerf, R.: La dynamique des solutions de macromolecules dans un champ de vitresses. Vol. 1, pp. 382–450.
Cesca, S., *Priola, A.* and *Bruzzone, M.*: Synthesis and Modification of Polymers Containing a System of Conjugated Double Bonds. Vol. 32, pp. 1–67.
Chiellini, E., *Solaro, R.*, *Galli, G.* and *Ledwith, A.*: Optically Active Synthetic Polymers Containing Pendant Carbazolyl Groups. Vol. 62, pp. 143–170.
Cicchetti, O.: Mechanisms of Oxidative Photodegradation and of UV Stabilization of Polyolefins. Vol. 7, pp. 70–112.
Clark, A. H. and *Ross-Murphy, S. B.*: Structural and Mechanical Properties of Biopolymer Gels. Vol. 83, pp. 57–193.
Clark, D.T.: ESCA Applied to Polymers. Vol. 24, pp. 125–188.
Cohen, R. E. see Argon, A. S.: Vol. 91/92, pp. 301–352.
Colemann, Jr., L. E. and *Meinhardt, N. A.*: Polymeriaztion Reactions of Vinyl Ketones. Vol. 1, pp. 159–179.
Comper, W. D. and *Preston, B. N.*: Rapid Polymer Transport in Concentrated Solutions. Vol. 55, pp. 105–152.
Corner, T.: Free Radical Polymerization – The Synthesis of Graft Copolymers. Vol. 62, pp. 95–142.
Cresceni, V.: Some Recent Studies of Polyelctrolyte Solutions. Vol. 5, pp. 358–386.
Crivello, J. V.: Cationic Polymerization – Iodonium and Sulfonium Salt Photoinitiators, Vol. 62, pp. 1–48.

Dave, R. see Kardos, J. L.: Vol. 80, pp. 101–123.
Davydov, B. E. and *Krentsel, B. A.*: Progress in the Chemistry of Polyconjugated Systems. Vol. 25, pp. 1–46.
Derouet, F. see Brosse, J.-C.: Vol. 81, pp. 167–224.
Dettenmaier, M.: Intrinsic Crazes in Polycarbonate Phenomenology and Molecular Interpretation of a New Phenomenon. Vol. 52/53, pp. 57–104.
Dettenmaier, M. and *Leberger, D.*: Crazing of Polymer-Diluent Mixtures. Vol. 91/92, pp. 119–136.
Diaz, A. F., *Rubinson, J. F.*, and *Mark, H. B., Jr.*: Electrochemistry and Electrode Application of Electroactive/Conductive Polymers. Vol. 84, pp. 113–140.
Dobb, M. G. and *McIntyre, J. E.*: Properties and Applications of Liquid-Crystalline Main-Chain Polymers. Vol. 60/61, pp. 61–98.
Döll, W.: Optical Interference Measurements and Fracture Mechanics Analysis of Crack Tip Craze Zones. Vol. 52/53, pp. 105–168.
Döll, W. and *Könczöl, L.*: Micromechanics of Fracture under Static and Fatigue Loading: Optical Interferometry of Crack Tip Craze Zones. Vol. 91/92, pp. 137–214.
Doi, Y. see *Keli, T.*: Vol. 73/74, pp. 201–248.
Dole, M.: Calorimetric of States and Transitions in Solid High Polymers. Vol. 2, pp. 221–274.
Donnet, J. B., *Vidal A.*: Carbon Black-Surface Properties and Interactions with Elastomers. Vol. 76, pp. 103–128.
Dorn, K., *Hupfer, B.*, and *Ringsdorf, H.*: Polymeric Monolayers and Liposomes as Models for Biomembranes How to Bridge the Gap Between Polymer Science and Membrame Biology? Vol. 64, pp. 1–54.
Dreyfuss, P. and *Dreyfuss, M. P.*: Polytetrahydrofuran. Vol. 4, pp. 528–590.
Drobnik, J. and *Rypáček, F.*: Soluble Synthetic Polymers in Biological Systems. Vol. 57, pp. 1–50.
Dröscher, M.: Solid State Extrusion of Semicrystalline Copolymers. Vol. 47, pp. 120–138.
Duduković, M. P. see Kardos, J. L.: Vol. 80, pp. 101–123.
Drzal, L. T.: The Interphase in Epoxy Composites. Vol. 75, pp. 1–32.
Dušek, K.: Network Formation in Curing of Epoxy Resins. Vol. 78, pp. 1–58.
Dušek, K. and *Prins. W.*: Structure and Elasticity of Non-Crystalline Polymer Networks. Vol. 6, pp. 1–102.
Dušek, K. see Brosse, J.-C.: Vol. 81, pp. 167–224.

Duncan, R. and *Kopeček, J.*: Soluble Synthetic Polymers as Potential Drug Carriers. Vol. 57, pp. 51–101.

Eastam, A. M.: Some Aspects of the Polymerization of Cyclic Ethers. Vol. 2, pp. 18–50.
Ehrlich, P. and *Mortimer, G. A.*: Fundamentals of the Free-Radical Polymerization of Ethylene. Vol. 7, pp. 386–448.
Eisenberg, A.: Ionic Forces in Polymers. Vol. 5, pp. 59–112.
Eiss, N. S. Jr. see Yorkgitis, E. M.: Vol. 72, pp. 79–110.
Elias, H.-G., Bareiss, R. und *Watterson, J. G.*: Mittelwerte des Molekulargewichts und anderer Eigenschaften. Vol. 11, pp. 111–204.
Elsner, G., Riekel, Ch. and *Zachmann, H. G.*: Synchrotron Radiation Physics. Vol. 67, pp. 1–58.
Elyashevich, G. K.: Thermodynamics and Kinetics of Orientational Crystallization of Flexible-Chain Polymers. Vol. 43, pp. 207–246.
Enikolopyan, N. S., Fridman, M. L., Stalnova, I. O. and *Popov, V. L.*: Filled Polymers: Mechanical Properties and Processability. Vol. 96, pp. 1–67.
Enikolopyan, N. S. see Ponomarenko, A. T.: Vol. 96, pp. 125–147.
Enkelmann, V.: Structural Aspects of the Topochemical Polymerization of Diacetylenes. Vol. 63, pp. 91–136.
Entelis, S. G., Evreinov, V. V. Gorshkov, A. V.: Functionally and Molecular Weight Distribution of Telchelic Polymers. Vol. 76, pp. 129–175.
Epaillard, F. see Brosse, J.-C.: Vol. 81, pp. 167–224.
Evreinov, V. V. see Entelis, S. G.: Vol. 76, pp. 129–175.

Ferruti, P. and *Barbucci, R.*: Linear Amino Polymers: Synthesis, Protonation and Complex Formation. Vol. 58, pp. 59–92.
Finkelmann, H. and *Rehage, G.*: Liquid Crystal Side-Chain Polymers. Vol. 60/61, pp. 99 bis 172.
Fischer, H.: Freie Radikale während der Polymerisation, nachgewiesen und identifiziert durch Elektronenspinresonanz. Vol. 5, pp. 463–530.
Flory, P. J.: Molecular Theory of Liquid Crystals. Vol. 59, pp. 1–36.
Ford, W. T. and *Tomoi, M.*: Polymer-Supported Phase Transfer Catalyst Reaction Mechanisms. Vol. 55, pp. 49–104.
Fradet, A. and *Maréchal, E.*: Kinetics and Mechansims of Polyesterifications. I. Reactions of Diols with Diacids. Vol. 43, pp. 51–144.
Franta, E. see Rempp, P.: Vol. 86, pp. 145–173.
Franz, G.: Polysaccharides in Pharmacy. Vol. 76, pp. 1–30.
Fridman, M. L. and *Servuk, V. D.*: Extension of Molten Polymers. Vol. 93, pp. 1–40.
Fridman, M. L. and *Peshkovsky, S. L.*: Molding of Polymers under Conditions of Vibration Effects. Vol. 93, pp. 41–80.
Fridman, M. L., Petrosyan, A.-Z., Levin, V. S. and *Bormashenko, E. Yu.*: Fundamentals of Low-Pressure Moulding of Polymer Pastes (Plastisols) and Thermoplastic Materials. Vol. 93, pp. 81–136.
Fridman, M. L. see Tunkel, V. I.: Vol. 93, pp. 137–174.
Fridman, M. L. see Sabsai, O. Yu.: Vol. 96, pp. 99–123.
Fridmann, M. L. see Enikolopyan, N. S.: Vol. 96, pp. 1–67.
Friedrich, K.: Crazes and Shear Bands in Semi-Crystalline Thermoplastics. Vol. 52/53, pp. 225–274.
Fujita, H.: Diffusion on Polymer-Diluent Systems. Vol. 3, pp. 1–47.
Funke, W.: Über die Strukturaufklärung vernetzter Makromoleküle, insbesondere vernetzter Polyesterharze, mit chemischen Methoden. Vol. 4, pp. 157–235.
Furukawa, H. see Kamon, T.: Vol. 80, pp. 173–202.

Gal'braikh, L. S. and *Rigovin, Z. A.*: Chemical Transformation of Cellulose. Vol. 14, pp. 87 bis 130.
Galli, G. see Chiellini, E.: Vol. 62, pp. 143–170.

Gallot, B. R. M.: Preparation and Study of Block Copolymers with Ordered Structures, Vol. 29, pp. 85–156.

Gandini, A.: The Behaviour of Furan Derivatives in Polymerization Reactions. Vol. 25, pp. 47–96.

Gandini, A. and *Cheradame, H.*: Cationic Polymerization. Initiation with Alkenyl Monomers. Vol. 34/35, pp. 1–289.

Geckeler, K., Pillai, V. N. R., and *Mutter, M.*: Applications of Soluble Polymeric Supports. Vol. 39, pp. 65–94.

Gerrens, H.: Kinetik der Emulsionspolymerisation. Vol. 1, pp. 234–328.

Ghiggino, K. P., Roberts, A. J. and *Phillips, D.*: Time-Resolved Fluorescence Techniques in Polymer and Biopolymer Studies. Vol. 40, pp. 69–167.

Gilham, J. K. see Aronhime, M. T.: Vol. 78, pp. 81–112.

Glöckner, G.: Analysis of Compositional and Structural Heterogeneitis of Polymer by Non-Exclusion HPCL. Vol. 79, pp. 159–214.

Godovsky, Y. K.: Thermomechanics of Polymers. Vol. 76, pp. 31–102.

Godovsky, Yu. K. and *Papkov, V. S.*: Thermotropic Mesophases in Element-Organic Polymers. Vol. 88, pp. 129–180.

Goethals, E. J.: The Formation of Cyclic Oligomers in the Cationic Polymerization of Heterocycles. Vol. 23, pp. 103–130.

Gorshkov, A. V. see Entelis, S. G.: Vol. 76, pp. 129–175.

Gräger, H. see Kulicke, W.-M.: Vol. 89, pp. 1–68.

Graessley, W. W.: The Etanglement Concept in Polymer Rheology. Vol. 16, pp. 1–179.

Graessley, W. W.: Entagled Linear, Branched and Network Polymer Systems. Molecular Theories. Vol. 47, pp. 67–117.

Grebowicz, J. see Wunderlich, B.: Vol. 60/61, pp. 1–60.

Grebowicz, J. see Wunderlich, B.: Vol. 87, pp. 1–121.

Greschner, G. S.: Phase Distribution Chromatography. Possibilities and Limitations. Vol. 73/74, pp. 1–62.

Hagihara, V., Sonogahira, K. and *Takahashi, S.*: Linear Polymers Containing Transition Metals in the Main Vol. 41, pp. 149–179.

Hallpap, P., Bölke, M., and *Heublein, G.*: Elucidation of Cationic Polymerization Mechanism by Means of Quantum Chemical Methods. Vol. 86, pp. 175–236.

Hasegawa, M.: Four-Center Photopolymerization in the Crystalline State. Vol. 42, pp. 1–49.

Hatano, M.: Induced Circular Dichroism in Biopolymer-Dye System. Vol. 77, pp. 1–121.

Hay, A. S.: Aromatic Polyethers. Vol. 4, pp. 496–527.

Hara, M. see Sauer, J. A.: Vol. 91/92, pp. 69–118.

Hayakawa, R. and *Wada, Y.*: Piezoelectricity and Related Properties of Polymer Films. Vol. 11, pp. 1–55.

Heidemann, E. and *Roth, W.*: Synthesis and Investigation of Collagen Model Peptides. Vol. 43, pp. 145–205.

Heinrich, G.; Straube, E., and *Helmis, G.*: Ruber Elasticity of Polymer Networks: Theories. Vol. 84, pp. 33–87.

Heitz, W.: Polymeric Reagents. Polymer Design, Scope, and Limitations. Vol. 23, pp. 1–23.

Helfferich, F.: Ionenaustausch. Vol. 1, pp. 329–381.

Helmis, G. see Heinrich, G.: Vol. 84, pp. 33–87.

Hendra, P. J.: Laser-Raman Spectra of Polymers. Vol. 6, pp. 151–169.

Hendrix, J.: Position Sensitive "X-ray Detectors". Vol. 67, pp. 59–98.

Henrici-Olivé, G. and *Olivé, S.*: Oligomerization of Ethylene with Soluble Transition-Metal Catalysts. pp. 496–577.

Henrici-Olivé, G. und *Olivé, S.*: Koordinative Polymerisation an löslichen Übergangsmetall-Katalysatoren. Vol. 6, pp. 421–472.

Henrici-Olivé, G. and *Olivé, S.*: Oligomerization of Ethylene with Soluble Transition-Metal Catalysts. Vol. 15, pp. 1–30.

Henrici-Olivé, G. and *Olivé, S.*: Molecular Interactions and Macroscopic Properties of Polyacrylonitrile and Model Substances. Vol. 32, pp. 123–152.

Henrici-Olivé, G. and *Olivé, S.*: The Chemistry of Carbon Fiber Formation from Polyacrylonitrile. Vol. 51, pp. 1–60.
Hermans, Jr., J., Lohr, D. and *Ferro, D.*: Treatment of the Folding and Unfolding of Protein Molecules in Solution According to a Lattic Model. Vol. 9, pp. 229–283.
Herz, J.-E. see Rempp, P.: Vol. 86, pp. 145–173.
Heublein, G. see Hallpap, P.: Vol. 86, pp. 175–236.
Higashimura, T. and *Sawamoto, M.*: Living Polymerization and Selective Dimerization: Two Extremes of the Polymer Synthesis by Cationic Polymerization. Vol. 62, pp. 49–94.
Higashimura, T. see Masuda, T.: Vol. 81, pp. 121–166.
Hlady, V. see Andrade, J. D.: Vol. 79, pp. 1–63.
Hoffman, A. S.: Ioninzing Radiation and Gas Plasma (or Glow) Discharge Treatments for Preparation of Novel Polymeric Biomaterials. Vol. 57, pp. 141–157.
Holzmüller, W.: Molecular Mobility, Deformation and Relaxation Processes in Polymers. Vol. 26, pp. 1–62.
Hori, Y. see Kashiwabara, H.: Vol. 82, pp. 141–207.
Horie, K. and *Mita, I.*: Reactions and Photodynamics in Polymer Solids. Vol. 88, pp. 77 bis 128.
Hutchinson, J. and *Ledwith, A.*: Photoinitiation of Vinyl Polymerization by Aromatic Carbonyl Compounds. Vol. 14, pp. 49–86.

Iizuka, E.: Properties of Liquid Crystals of Polypeptides: with Stress on the Electromagnetic Orientation. Vol. 20, pp. 79–107.
Ikada, Y.: Characterization of Graft Copolymers. Vol. 29, pp. 47–84.
Ikada, Y.: Blood-Compatible Polymers. Vol. 57, pp. 103–140.
Ikada, Y. see Tabata, Y.: Vol. 94, pp. 107–142.
Imanishi, Y.: Synthese, Conformation, and Reactions of Cyclic Peptides. Vol. 20, pp. 1–77.
Inagaki, H.: Polymer Separation and Characterization by Thin-Layer Chromatography. Vol. 24, pp. 189–237.
Inoue, S.: Asymmetric Reactions of Synthetic Polypeptides. Vol. 21, pp. 77–106.
Irie, M.: Photoresponsive Polymers. Vol. 94, pp. 27–68.
Ise, N.: Polymerizations under an Electric Field. Vol. 6, pp. 347–376.
Ise, N.: The Mean Activity Coefficient of Polyelectrolytes in Aqueous Solutions and Its Related Properties. Vol. 7, pp. 536–593.
Isihara, A.: Irreversible Processes in Solutions of Chain Polymers. Vol. 5, pp. 531–567.
Isihara, A.: Intramolecular Statistics of a Flexible Chain Molecule. Vol. 7, pp. 449–476.
Isihara, A. and *Guth, E.*: Theory of Dilute Macromolecular Solutions. Vol. 5, pp. 233–260.
Ishikawa, M. see Narisawa, I.: Vol. 91/92, pp. 353–392.
Iwatsuki, S.: Polymerization of Quinodimethane Compounds. Vol. 58, pp. 93–120.

Janeschitz-Kriegl, H.: Flow Birefrigence of Elastico-Viscous Polymer Systems. Vol. 6, pp. 170–318.
Jenkins, R. and *Porter, R. S.*: Unpertubed Dimensions of Stereoregular Polymers. Vol. 36, pp. 1–20.
Jenngins, B. R.: Electro-Optic Methods for Characterizing Macromolecules in Dilute Solution. Vol. 22, pp. 61–81.
Johnston, D. S.: Macrozwitterion Polymerization. Isihara, A.: Vol. 42, pp. 51–106.

Kamachi, M.: Influence of Solvent on Free Radical Polymerization of Vinyl Compounds. Vol. 38, pp. 55–87.
Kamachi, M.: ESR Studies on Radical Polymerization. Vol. 82, pp. 207–277.
Kamide, K. and *Saito, M.*: Cellulose and Cellulose Derivatives: Recent Advances in Physical Chemistry. Vol. 83, pp. 1–57.
Kamon, T., Furukawa, H.: Curing Mechanisms and Mechanical Properties of Cured Epoxy Resins. Vol. 80, pp. 173–202.
Kaneda, A. see Kinjo, N.: Vol. 88, pp. 1–48.

Kaneko, M. and *Wöhrle, D.:* Polymer-Coated Electrodes: New Materials for Science and Industry. Vol. 84, pp. 141–228.
Kaneko, M. and *Yamada, A.:* Solar Energy Conversion by Functional Polymers. Vol. 55, pp. 1–48.
Kardos, J. L., Duduković, M. P., Dave, R.: Void Growth and Resin Transport During Processing of Thermosetting – Matrix Composits. Vol. 80, pp. 101–123.
Kashiwabra, H., Shimada, S., Hori, Y. and *Sakaguchi, M.:* ESR Application to Polymer Physics – Molecular Motion in Solid Matrix in which Free Radicals are Trapped. Vol. 82, pp. 141–207.
Kawabata, S. and *Kawai, H.:* Strain Energy Density Functions of Rubber Vulcanizates from Biaxial Extension. Vol. 24, pp. 89–124.
Keli, T., Doi, Y.: Synthesis of "Living" Polyolefins with Soluble Ziegler-Natta Catalysts and Application to Block Copolymerization. Vol. 73/74, pp. 201–248.
Kelly, F. N. see LeMay, J. D.: Vol. 78, pp. 113–148.
Kennedy, J. P. and *Chou, T.:* Poly(isobutylene-*co*-β-Pinene): A New Sulfur Vulcanizable, Ozone Resistant Elastomer by Cationic Isomerization Copolymerization. Vol. 21, pp. 1–39.
Kennedy, J. P. and *Delvaux, J. M.:* Synthesis, Characterization and Morphology of Poly (butadieneg-Styrene). Vol. 38, pp. 141–163.
Kennedy, J. P. and *Gillham, J. K.:* Cationic Polymerization of Olefins with Alkylaluminium Initiators. Vol. 10, pp. 1–33.
Kennedy, J. P. and *Johnston, J. E.:* The Cationic Isomerization Polymerization of 3-Methyl-1-butene and 4-Methyl-1-pentene. Vol. 19, pp. 57–95.
Kennedy, J. P. and *Langer, Jr., A. W.:* Recent Advances in Cationic Polymerization. Vol. 3, pp. 508–580.
Kennedy, J. P. and *Otsu, T.:* Polymerization with Isomerization of Monomer Preceding Propagation. Vol. 7, pp. 369–385.
Kennedy, J. P. and *Rengachary, S.:* Correlation Between Cationic Model and Polymerization Reactions of Olefins. Vol. 14, pp. 1–48.
Kennedy, J. P. and *Trivedi, P. D.:* Cationic Olefin Polymerization Using Alkyl Halide – Alkyl-Aluminium Initiator Systems. I. Reactivity Studies. II. Molecular Weight Studies. Vol. 28, pp. 83–151.
Kennedy, J. P., Chang, V. S. and *Guyot, A.:* Carbocationic Synthesis and Characterization of Polyolefins with Si-H and Si-Cl Head Groups. Vol. 43, pp. 1–50.
Khoklov, A. R. and *Grosberg, A. Yu.:* Statistical Theory of Polymeric Lyotropic Liquid Crystals. Vol. 41, pp. 53–97.
Kinjo, N., Ogata, M., Nishi, K. and *Kaneda, A.:* Epoxy Molding Compounds as Encapsulation Materials for Microelectronic Devices. Vol. 88, pp. 1–48.
Kinloch, A. J.: Mechanics and Mechanics of Fracture of Thermosetting Epoxy Polymers. Vol. 72, pp. 45–68.
Kissin, Yu. V.: Structures of Copolymers of High Olefins. Vol. 15, pp. 91–155.
Kitagawa, T. and *Miyazawa, T.:* Neutron Scattering and Normal Vibrations of Polymers. Vol. 9, pp. 335–414.
Kitamaru, R. and *Horii, F.:* NMR Approach to the Phase Structure of Linear Polyethylene. Vol. 26, pp. 139–180.
Klosinski, P., Penczek, S.: Teichoic Acids and Their Models: Membrane Biopolymers with Polphosphate Backbones. Synthesis, Structure and Properties. Vol. 79, pp. 139–157.
Kloosterboer, J. G.: Network Formation by Chain Crosslinking Photopolymerization and its Applications in Electronics. Vol. 84, pp. 1–62.
Knappe, W.: Wärmeleitung in Polymeren. Vol. 7, pp. 477–535.
Koenik, J. L. see Mertzel, E. Vol. 75, pp. 73–112.
Koenig, J. L.: Fourier Transforms Infrared Spectroscopy of Polymers. Vol. 54, pp. 87–154.
Koenig, J. L. see Andreis, M. Vol. 89, pp. 69–160.
Könczöl, L. see Döll, W.: Vol. 91/92, pp. 137–214.
Kötter, M. see Kulicke, W.-M. Vol. 89, pp. 1–68.

Kolařik, J.: Secondary Relaxations in Glassy Polymers: Hydrophilic Polymethacrylates and Polyacrylates: Vol. 46, pp. 119–161.
Kong, E. S. W.: Physical Aging in Epoxy Matrices and Composites. Vol. 80, pp. 125–171.
Koningsveld, R.: Preparative and Analytical Aspects of Polymer Fractionation. Vol. 7.
Kosyanchuk, L. F. see Lipatov, Yu. S.: Vol. 88, pp. 49–76.
Kothe, G. see Müller, K.: Vol. 95, pp. 1–56.
Kovacs, A. J.: Transition vitreuse dans les polymers amorphes. Etude phénoménologique. Vol. 3, pp. 394–507.
Krässig, H. A.: Graft Co-Polymerization of Cellulose and Its Derivates. Vol. 4, pp. III–156.
Kramer, E. J.: Microscopic and Molecular Fundamentals of Crazing. Vol. 52/53, pp. 1–56.
Kramer, E. J. and *Berger, L. L.:* Fundamental Processes of Craze Growth and Fracture. Vol. 91/92, pp. 1–68.
Kraus, G.: Reinforcement of Elastomers by Carbon Black. Vol. 8, pp. 155–237.
Kratochvila, J. see Mejzlik, J.: Vol. 81, pp. 83–120.
Kreutz, W. and *Welte, W.:* A General Theory for the Evaluation of X-Ray Diagrams of Biomembranes and Other Lamellar Systems. Vol. 30, pp. 161–225.
Krimm, S.: Infrared Spectra of High Polymers. Vol. 2, pp. 51–72.
Kuhn, W., Ramel, A., Walters, D. H. Ebner, G. and *Kuhn, H. J.:* The Production of Mechanical Energy from Different Forms of Chemical Energy with Homogeneous and Cross-Striated High Polymer Systems. Vol. 1, pp. 540–592.
Kulicke, W.-M., Kötter, M. and *Gräger, H.:* Drag Reduction Phenomenon with Special Emphasis on Homogeneous Polymer Solutions. Vol. 89, pp. 1–68.
Kunitake, T. and *Okahata, Y.:* Catalytic Hydrolysis by Synthetic Polymers. Vol. 20, pp. 159 to 221.
Kurata, M. and *Stockmayer, W. H.:* Intrinsic Viscosities and Unperturbed Dimensions of Long Chain Molecules. Vol. 3, pp. 196–312.
Kurijama, I. see Nakase, Y.: Vol. 65, pp. 79–134.
Kurimura, Y.: Macromolecule-Metal Complexes – Reactions and Molecular Recognition. Vol. 90, pp. 105–138.

Lazar, M., Rychly, I., and *Rado, R.:* Crosslinking of Polyolefins. Vol. 95, pp. 149–198.
Leberger, D. see Dettenmaier, M.: Vol. 65, pp. 79–134.
Ledwith, A. and *Sherrington, D. C.:* Stable Organic Cation Salts: Ion Pair Equilibria and Use in Cationic Polymerization. Vol. 19, pp. 1–56.
Ledwith, A. see Chiellini, E. Vol. 62, pp. 143–170.
Lee, C.-D. S. and *Daly, W. H.:* Mercaptan-Containing Polymers. Vol. 15, pp. 61–90.
Legeay, G. see Brosse, J.-C.: Vol. 81, pp. 167–224.
LeMay, J. D., Kelly, F. N.: Structure and Ultimate Properties of Epoxy Resins. Vol. 78, pp. 113–148.
Lesná, M. see Mejzlik, J.: Vol. 81, pp. 83–120.
Levin, V. S. see Fridman, M. L.: Vol. 93, pp. 81–136.
Lindberg, J. J. and *Hortling, B.:* Cross Polarization – Magic Angle Spinning NMR Studies of Carbohydrates and Aromatic Polymers. Vol. 66, pp. 1–22.
Lipatov, Y. S.: Relacation and Viscoelastic Properties of Heterogeneous Polymeric Compositions. Vol. 22, pp. 1–59.
Lipatov, Y. S.: The Iso-Free-Volume State and Glass Transition in Amorphous Polymers: New Development of the Theory. Vol. 26, pp. 63–104.
Lipatov, Yu. S., Lipatova, T. E. and *Kosyanchuk, L. F.:* Synthesis and Structure of Macromolecular Topological Compounds. Vol. 88, pp. 49–76.
Lipatova, T. E.: Medical Polymer Adhesives. Vol. 79, pp. 65–93.
Lipatova, T. E. see Lipatov, Yu. S.: Vol. 88, pp. 49–76.
Litmanovich, A. A. see Papisov, J. M.: Vol. 90, pp. 139–180.
Lohse, F., Zweifel, H.: Photocrosslinking of Epoxy Resins. Vol. 78, pp. 59–80.
Lustoň, J. and *Vašš, F.:* Anionic Copolymerization of Cyclic Ethers with Cyclic Anhydrides. Vol. 56, pp. 91–133.

Madec, J.-P. and *Maréchal, E.*: Kinetics and Mechanisms of Polyesterifications. II. Reactions of Diacids with Diepoxides. Vol. 71, pp. 153–228.
Malkin, A. Ya. and *Zhirkov, P. V.*: Flow of Polymerizing Liquids. Vol. 95, pp. 111–148.
Malkin, A. Ya.: Rheology of Filled Polymers. Vol. 96, pp. 69–97.
Mano, E. B. and *Coutinho, F. M. B.*: Grafting on Polyamides. Vol. 19, pp. 97–116.
Maréchal, E. see Madec, J.-P. Vol. 71, pp. 153–228.
Mark, H. B., Jr. see Diaz, A. F.: Vol. 84, pp. 113–140.
Mark, J. E.: The Use of Model Polymer Networks to Elucidate Molecular Aspects of Rubberlike Elasticity. Vol. 44, pp. 1–26.
Mark, J.E. see Queslel, J. P. Vol. 71, pp. 229–248.
Maser, F., Bode, K., Pillai, V. N. R. and *Mutter, M.*: Conformational Studies on Model Peptides. Their Contribution to Synthetic, Structural and Functional Innovations on Proteins. Vol. 65, pp. 177–214.
Masuda, T. and *Higashimura, T.*: Polyacetylenes with Substituents: Their Synthesis and Properties. Vol. 81, pp. 121–166.
McGrath, J. E. see Yilgör, I.: Vol. 86, pp. 1–86.
McGrath, J. E. see Yorkgitis, E. M. Vol. 72, pp. 79–110.
McIntyre, J. E. see Dobb, M. G. Vol. 60/61, pp. 61–98.
Meerwall v., E. D.: Self-Diffusion in Polymer Systems. Measured with Field-Gradient Spin Echo NMR Methods, Vol. 54, pp. 1–29.
Mejzlik, J., Lesná, M. and *Kratochvila, J.*: Determination of the Number of Active Centers in Ziegler-Natta Polymerizations of Olefins. Vol. 81, pp. 83–120.
Mengoli, G.: Feasibility of Polymer Film Coating Through Electrointiated Polymerization in Aqueous Medium. Vol. 33, pp. 1–31.
Mertzel, E., Koenik, J. L.: Application of FT-IR and NMR to Epoxy Resins. Vol. 75, pp. 73–112.
Meyerhoff, G.: Die viscosimetrische Molekulargewichtsbestimmung von Polymeren. Vol. 3, pp. 59–105.
Millich, F.: Rigid Rods and the Characterization of Polysocyanides. Vol. 19, pp. 117–141.
Mita, I. see Horie, K.: Vol. 88, pp. 77–128.
Möller, M.: Cross Polarization – Magie Angle Sample Spinning NMR Studies. With Respect to the Rotational Isomeric States of Saturated Chain Molecules. Vol. 66, pp. 59–80.
Möller, M. see Wunderlich, B.: Vol. 87, pp. 1–121.
Morawetz, H.: Specific Ion Binding by Polyelectrolytes. Vol. 1, pp. 1–34.
Morgan, R. J.: Structure-Property Relations of Epoxies Used as Composite Matrices. Vol. 72, pp. 1–44.
Morin, B. P., Breusova, I. P. and *Rogovin, Z. A.*: Structural and Chemical Modification of Cellulose by Graft Copolymerization. Vol. 42, pp. 139–166.
Müller, K., Kothe, G., and *Wassmer, K.-H.*: Dynamic Magnetic Resonance of Liquid crystal Polymers: Molecular Organization and Macroscopic Properties. Vol. 95, pp. 1–56.
Mulvaney, J. E., Oversberger, C. C. and *Schiller, A. M.*: Anionic Polymerization. Vol. 3, pp. 106–138.

Nakase, Y., Kurijama, I. and *Odajima, A.*: Analysis of the Fine Structure of Poly(Oxymethylene) Prepared by Radiation-Induced Polymerization in the Solid State. Vol. 65, pp. 79–134.
Narisawa, I. and *Ishikawa, M.*: Crazing in Semicrystalline Thermoplastics. Vol. 91/92, pp. 353–392.
Neuse, E.: Aromatic Polybenzimidazoles. Syntheses, Properties, and Applications. Vol. 47, pp. 1–42.
Nicolais, L. see Apicella, A. Vol. 72, pp. 69–78.
Nikolaeva, N. E. see Sabsai, O. Yu.: Vol. 96, pp. 99–123.
Nishi, K. see Kinjo, N.: Vol. 88, pp. 1–48.

Noristi, L. see Barbé, P. C.: Vol. 81, pp. 1–83.
Nuyken, O., *Weidner, R.*: Graft and Block Copolymers via Polymeric Azo Initiators. Vol. 73/74, pp. 145–200.

Ober, Ch. K., *Jin, J.-J.* and *Lenz, R. W.*: Liquid Crystal Polymers with Flexible Spacers in the Main Chain. Vol. 59, pp. 103–146.
Odajima, A. see Nakase, Y.: Vol. 65, pp. 79–134.
Ogata, M. see Kinjo, N.: Vol. 88, pp. 1–48.
Okubo, T. and *Ise, N.*: Synthetic Polyelectrolytes as Models of Nucleic Acids and Esterases. Vol. 25, pp. 135–181.
Oleinik, E. F.: Epoxy-Aromatic Amine Networks in the Classy State Structure and Properties. Vol. 80, pp. 49–99.
Osaki, K.: Viscoelastic Properties of Dilute Polymer Solutions. Vol. 12, pp. 1–64.
Osada, Y.: Conversion of Chemical Into Mechanical Energy by Synthetic Polymers (Chemomechanical Systems). Vol. 82, pp. 1–47.
Osada, Y. and *Biedermann, H.*: Plasma Chemistry of Polymers. Vol. 95, pp. 57–110.
Oster, G. and *Nishijima, Y.*: Fluorescence Methods in Polymer Science. Vol. 3, pp. 313–331.
Otsu, T. see Sato, T. Vol. 71, pp. 41–78.
Overberger, C. G. and *Moore, J. A.*: Ladder Polymers. Vol. 7, pp. 113–150.
Oversberger, C. C. see Mulvaney, J. E.: Vol. 3, pp. 106–138.

Packirisamy, S. see Biswas, M. Vol. 70, pp. 71–118.
Papisov, J. M. and *Litmanovich, A. A.*: Molecular „Recognition" in Interpolymer Interactions and Matrix Polymerization. Vol. 90, pp. 139–180.
Papkov, S. P.: Liquid Crystalline Order in Solutions of Rigid-Chain Polymers. Vol. 59, pp. 75–102.
Papkov, V. S. see Godovsky, Yu. K.: Vol. 88, pp. 129–180.
Patrat, F, *Killmann, E.* und *Schiebener, C.*: Die Absorption von Makromolekülen aus Lösung. Vol. 3, pp. 332–393.
Patterson, G. D.: Photon Correlation Spectroscopy of Bulk Polymers. Vol. 48, pp. 125–159.
Pencek, S., *Kubisa, P.* and *Matyjaszewski, K.*: Cationic Ring-Opening Polymerization of Heterocyclic Monomers. Vol. 37, pp. 1–149.
Pencek, S., *Kubisa, P.* and *Matyjaszewski, K.*: Cationic Ring-Opening Polymerization; 2. Synthetic Applications. Vol. 68/69, pp. 1–298.
Penczek, S. see Klosinski, P.: Vol. 79, pp. 139–157.
Peshkovsky, S. L. see Fridman, M. L.: Vol. 93, pp. 41–80.
Peticolas, W. L.: Inelastic Laser Light Scattering from Biological and Synthetic Polymers. Vol. 9, pp. 285–333.
Petropoulos, J. H.: Membranes with Non-Homogeneous Sorption Properties. Vol. 64, pp. 85–134.
Petrosyan, A. Z. see Fridmann, M. L.: Vol. 93, pp. 81–136.
Pino, P.: Optically Active Addition Polymers. Vol. 4, pp. 393–456.
Pitha, J.: Physiological Activities of Synthetic Analogs of Polynucleotides. Vol. 50, pp. 1 to 16.
Platé, N. A. and *Noak, O. V.*: A Theoretical Consideration of the Kinetics and Statistics of Reactions of Functional Groups of Macromolecules. Vol. 79, pp. 95–138.
Platé, N. A. see Shibaev, V. P. Vol. 60/61, pp. 173–252.
Plesch, P. H.: The Propagation Rate-Constants in Cationic Polymerisations. Vol. 8, pp. 137 to 154.
Ponomarenko, A. T., *Shevchenko, V. G.* and *Enikolopyan, N. S.*: Formation Processes and Properties of Conducting Polymer Composites. Vol. 96, pp. 125–147.
Popov, V. L. see Enikolopyan, N. S.: Vol. 96, pp. 1–67.
Porod, G.: Anwendung und Ergebnisse der Röntgenkleinwinkelstreuung in festen Hochpolymeren. Vol. 2, pp. 363–400.
Pospišil, J.: Transformations of Phenolic Antioxidants and the Role of Their Products in the Long-Term Properties of Polyolefins. Vol. 36, pp. 69–133.

Postelnek, W., Colemann, L. E., and *Lovelace, A. M.:* Fluorine-Containing Polymers. I. Fluorinated Vinyl Polymers with Functional Groups, Condensation Polymers, and Styrene Polymers. Vol. 1, pp. 75–113.

Queslel, J. P. and *Mark, J. E.:* Molecular Interpretation of the Moduli Elastomeric Polymer Networks of Know Structure. Vol. 65, pp. 135–176.
Queslel, J. P. and *Mark, J. E.:* Swelling Equilibrium Studies of Elastomeric Network Structures. Vol. 71, pp. 229–248.

Rado, R. see Lazar, M.: Vol. 95, pp. 149–198.
Rehage, G. see Finkelmann, H. Vol. 60/61, pp. 99–172.
Rempp, P. F. and *Franta, E.:* Macromonomers:Synthesis, Characterization and Applications. Vol. 58, pp. 1–54.
Rempp, P., Herz, J. and *Borchard, W.:* Model Networks. Vol. 26, pp. 107–137.
Rempp, P., Franta, E., and *Herz, J.-E.:* Macromolecular Engineering by Anionic Methods. Vol. 86, pp. 145–173.
Richards, R. W.: Small Angle Neutron Scattering from Block Copolymers. Vol. 71, pp. 1 to 40.
Rigbi, Z.: Reinforcement of Rubber by Carbon Black. Vol. 36, pp. 21–68.
Rigbi, D. see Roe, R.-J.: Vol. 82, pp. 103–141.
Roe, R.-J. and *Rigby, D.:* Phase Relations and Miscibility in Polymer Blends Containing Copolymers. Vol. 82, pp. 103–141.
Rogovin, Z. A. and *Gabrielyan, G. A.:* Chemical Modifications of Fibre Forming Polymers and Copolymers of Acrylonitrile. Vol. 25, pp. 97–134.
Roha, M.: Ionic Factors in Steric Control. Vol. 4, pp. 353–392.
Roha, M.: The Chemistry of Coordinate Polymerization of Dienes. Vol. 1, pp. 512–539.
Ross-Murphy, S. B. see Clark, A. H.: Vol. 83, pp. 57–193.
Rostami, S. see Walsh, D. J. Vol. 70, pp. 119–170.
Rozengerk, v. A.: Linetics, Thermodynamics and Mechanism of Reactions of Epoxy Oligomers with Amines. Vol. 75, pp. 113–166.
Rubinson, J. F. see Diaz, A. F.: Vol. 84, pp. 113–140.
Rychly, I. see Lazar, M.: Vol. 95, pp. 149–198.

Sabsai, O. Yu., Nikolaeva, N. E. and Fridman, M. L.: Rheology of Gas-Containing Polymer Systems. Vol. 96, pp. 99–123.
Safford, G. J. and *Naumann, A. W.:* Low Frequency Motions in Polymers as Measured by Neutron Inelastic Scattering. Vol. 5, pp. 1–27.
Sakaguchi, M. see Kashiwabara, H.: Vol. 82, pp. 141–207.
Saito, M. see Kamide, K.: Vol. 83, pp. 1–57.
Sato, T. and *Otsu, T.:* Formation of Living Propagating Radicals in Microspheres and Their Use in the Synthesis of Block Copolymers. Vol. 71, pp. 41–78.
Sauer, J. A. and *Chen, C. C.:* Crazing and Fatigue Behavior in One and Two Phase Glassy Polymers. Vol. 52/53, pp. 169–224.
Sauer, J. A. and *Hara, M.:* Effect of Molecular Variables on Crazing and Fatigue of Polymers. Vol. 91/92, pp. 69–118.
Sawamoto, M. see Higashimura, T. Vol. 62, pp. 49–94.
Schiller, A. M. see Mulvaney, J. E.: Vol. 3, pp. 106–138.
Schirrer, R.: Optical Interferometry: Running Crack-Tip Morphologies and Craze Material Properties. Vol. 91/92, pp. 215–262.
Schmidt, R. G., Bell, J. P.: Epoxy Adhesion to Metals. Vol. 75, pp. 33–72.
Schuerch, C.: The Chemical Synthesis and Properties of Polysaccarides of Biomedical Interest. Vol. 10, pp. 173–194.
Schulz, R. C. und *Kaiser, E.:* Synthese und Eigenschaften von optische aktiven Polymeren. Vol. 4, pp. 236–315.
Seanor, D. A.: Charge Transfer in Polymers. Vol. 4, pp. 317–352.

Semerak, S. N. and *Frank, C. W.*: Photophysics of Excimer Formation in Aryl Vinyl Polymers, Vol. 54, pp. 31–85.
Seidl, J., Malinský, J., Dušek, K. und *Heitz, W.*: Makroporöse Styrol-Divinylbenzol-Copolymere und ihre Verwendung in der Chromatographie und zur Darstellung von Ionenaustauschen. Vol. 5, pp. 113–213.
Semjonow, V.: Schmelzviskositäten hochpolymerer Stoffe. Vol. 5, pp. 387–450.
Semlyen, J. A.: Ring-Chain Equilibria and the Conformations of Polymer Chains. Vol. 21, pp. 41–75.
Sen, A.: The Copolymerization of Carbon Monoxide with Olefins. Vol. 73/74, pp. 125–144.
Senturia, S. D., Sheppard, N. F. Jr.: Dielectric Analysis of Thermoset Cure. Vol. 80, pp. 1 to 47.
Sevruk, V. D. see Fridmann, M. L.: Vol. 93, pp. 1–40.
Sharkey, W. H.: Polymerizations Through the Carbon-Sulphur Double Bond. Vol. 17, pp. 73–103.
Sheppard, N. F. Jr. see Senturia, S. D.: Vol. 80, pp. 1–47.
Shevchenko, V. G. see Ponomarenko, A. T.: Vol. 96, pp. 125–147.
Shibaev, V. P. and *Platé, N. A.*: Thermotropic Liquid-Crystalline Polymers with Mesogenic Side Groups, Vol. 60/61, pp. 173–252.
Shimada, S. see Kashiwabara, H.: Vol. 82, pp. 141–207.
Shimidzu, T.: Cooperative Actions in the Nucleophile-Containing Polymers. Vol. 23, pp. 55 to 102.
Shutov, F. A.: Foamed Polymers Based on Reactive Oligomers, Vol. 39, pp. 1–64.
Shutov, F. A.: Foamed Polymers. Cellular Structure and Properties. Vol. 51, pp. 155–218.
Shutov, F. A.: Syntactic Polymer Foams. Vol. 73/74, pp. 63–124.
Siesler, H. W.: Rheo-Optical Fourier-Tranform Infrared Spectroscopy: Vibrational Spěctra and Mechanical Properties of Polymers. Vol. 65, pp. 1–78.
Silvestri, G., Gambino, S., and *Filardi, G.*: Electrochemical Production of Initiators for Polymerization Processes. Vol. 38, pp. 27–54.
Sixl, H.: Spectroscopy of the Intermediate States of the Solid State Polymerization Reaction in Diacetylene Crystals. Vol. 63, pp. 49–90.
Slichter, W. P.: The Study of High Polymers by Nuclear Magnetic Resonance. Vol. 1, pp. 35–74.
Small, P. A.: Long-Chain Branching in Polymers. Vol. 18.
Smets, G.: Block and Graft Copolymers. Vol. 2, pp. 173–220.
Smets, G.: Photochromic Phenomena in the Solid Phase. Vol. 50, pp. 17–44.
Smets, G.: see Beylen, M. van: Vol. 86, pp. 87–143.
Sohma, J. and *Sakaguchi, M.*: ESR Studies on Polymer Radicals Produced by Mechanical Destruction and Their Reactivity. Vol. 20, pp. 109–158.
Solaro, R. see Chiellini, E. Vol. 62, pp. 143–170.
Sotobayashi, H. und *Springer, J.*: Oligomere in verdünnten Lösungen. Vol. 6, pp. 473–548.
Soutif, J.-C. see Brosse, J.-C.: Vol. 81, pp. 167–224.
Sperati, C. A. and *Starkweather, Jr., H. W.*: Fluorine-Containing Polymers. II. Polytetrafluoroethylene. Vol. 2, pp. 465–495.
Spiertz, E. J. see Vollenbroek, F. A.: Vol. 84, pp. 85–112.
Spiess, H. W.: Deuteron NMR – A new Toolfor Studying Chain Mobility and Orientation in Polymers. Vol. 66, pp. 23–58.
Sprung, M. M.: Recent Progress in Silicone Chemistry. I. Hydrolysis of Reactive Silane Intermediates, Vol. 2, pp. 442–464.
Stahl, E. and *Brüderle, V.*: Polymer Analysis by Thermofractography. Vol. 30, pp. 1–88.
Stalnova, I. O. see Enikolopyan, N. S.: Vol. 96, pp. 1–67.
Stannett, V. T., Koros, W. J., Paul, D. R., Lonsdale, H. K., and *Baker, R. W.*: Recent Advances in Membrane Science and Technology. Vol. 32, pp. 69–121.
Stavermann, A. J.: Properties of Phantom Networks and Real Networks. Vol. 44, pp. 73 to 102.
Stauffer, D., Coniglio, A. and *Adam, M.*: Gelation and Critical Phenomena. Vol. 44, pp. 103–158.

Stille, J. K.: Diels-Adler Polymerization. Vol. 3, pp. 48–58.
Stolka, M. and *Pai, D.:* Polymers with Photoconductive Properties. Vol. 29, pp. 1–45.
Straube, E. see Heinrich, G.: Vol. 84, pp. 33–87.
Stuhrmann, H.: Resoance Scattering in Macromolecular Structure Research. Vol. 67, pp. 123–164.
Subramanian, R. V.: Electroinitiated Polymerization on Electrodes. Vol. 33, pp. 35–58.
Sumitomo, H. and *Hashimoto, K.:* Polyamides as Barrier Materials. Vol. 64, pp. 55–84.
Sumitomo, H. and *Okada, M.:* Ring-Opering Polymerization of Bicyclic Acetals, Oxalactone, and Oxalactam. Vol. 28, pp. 47–82.
Szegö, L.: Modified Polyethylene Terephthalate Fibers. Vol. 31, pp. 89–131.
Swarc, M.: Termination of Anionic Polymerization. Vol. 2, pp. 275–306.
Swarc, M.: The Kinetics and Mechanism of N-carboxy-α-amino-acid Anhydride (NCA) Polymerization to Poly-amino Acids. Vol. 4, pp. 1–65.
Szwarc, M.: Thermodynamics of Polymerization with Special Emphasis on Living Polymers. Vol. 4, pp. 457–495.
Szwarc, M.: Living Polymers and Mechanisms of Anionic Polymerization. Vol. 49, pp. 1 to 175.
Swarc, M. see Beylen, M. van: Vol. 86, pp. 87–143.

Tabata, Y. and *Ikada, Y.:* Phagocytosis of Polymer Microspheres by Macrophages. Vol. 94, pp. 107–142.
Takahashi, A. and *Kawaguchi, M.:* The Structure of Macromolecules Adsorbed on Interfaces. Vol. 46, pp. 1–65.
Takekoshi, T.: Polyimides. Vol. 94, pp. 1–26.
Takemori, M. T.: Competition Between Crazing and Shear Flow During Fatigue. Vol. 91/92, pp. 263–300.
Takemoto, K. and *Inaki, Y.:* Synthetic Nuclei Acid Analogs. Preparation and Interactions. Vol. 41, pp. 1–51.
Tani, H.: Stereospecific Polymerization of Aldehydes and Epoxides. Vol. 11, pp. 57–110.
Tate, B. E.: Polymerization of Itaconic Acid and Derivatives. Vol. 5, pp. 214–232.
Tazuke, S.: Photosensitized Charge Transfer Polymerization. Vol. 6, pp. 321–346.
Teramoto, A. and *Fujita, H.:* Conformation-dependet Properties of Synthetic Polypeptides in the Helix-Coil Transition Region. Vol. 18, pp. 65–149.
Theocaris, P. S.: The Mesophase and its Influence on the Mechanical Behvior of Composites. Vol. 66, pp. 149–188.
Thomas, W. M.: Mechanismus of Acrylonitrile Polymerization. Vol. 2, pp. 401–441.
Tieke, B.: Polymerization of Butadiene and Butadiyne (Diacetylene) Derivatives in Layer Structures. Vol. 71, pp. 79–152.
Tobolsky, A. V. and *DuPré, D. B.:* Macromolecular Relaxation in the Damped Torsional Oscillator and Statistical Segment Models. Vol. 6, pp. 103–127.
Tosi, C. and *Ciampelli, F.:* Applications of Infrared Spectroscopy to Ethylene-Propylene Copolymers. Vol. 12, pp. 87–130.
Tosi, C.: Sequence Distribution in Copolymers: Numerical Tables. Vol. 5, pp. 451–462.
Tran, C. see Yorgitis, E. M. Vol. 72, pp. 79–110.
Tsuchida, E. and *Nishide, H.:* Polymer-Metal Complexes and Their Catalytic Activity. Vol. 24, pp. 1–87.
Tsuji, K.: ESR Study of Photodegradation of Polymers. Vol. 12, pp. 131–190.
Tsvetkov, V. and *Andreeva, L.:* Flow and Electric Birefringence in Rigid-Chain Polymer Solutions. Vol. 39, pp. 95–207.
Tunkel, V. I. and *Fridman, M. L.:* Granulated Thermosetting Materials (Aminoplasts) – Technology. Vol. 93, pp. 137–174.
Tuzar, Z.; Kratochvill, P., and *Bohdanecký, M.:* Dilute Solution Properties of Aliphatic Polyamides. Vol. 30, pp. 117–159.

Uematsu, I. and *Uematsu, Y.:* Polypeptide Liquid Crystals. Vol. 59, pp. 37–74.

Valuev, L. I. see Platé, N. A.: Vol. 79, pp. 95–138.
Valvassori, A. and *Sartori, G.*: Present Status of the Multicomponent Copolymerization Theory. Vol. 5, pp. 28–58.
Vidal, A. see Donnet, J. B. Vol. 76, pp. 103–128.
Viovy, J. L. and *Monnerie, L.*: Fluorescence Anisotropy Technique Using Synchroton Radiation as a Powerful Means for Studying the Orientation Correlation Functions of Polymer Chains. Vol. 67, pp. 99–122.
Voigt-Martin, I.: Use of Transmission Electron Microscopy to Obtain Quantitative Information About Polymers. Vol. 67, pp. 195–218.
Vollenbrock, F. A. and *Spiertz, E. J.*: Photoresist Systems for Microlithography. Vol. 84, pp. 85–112.
Voorn, M. J.: Phase Separation in Polymer Solutions. Vol. 1, pp. 192–233.

Walsh, D. J., Rostami, S.: The Miscibility of High Polymers: The Role of Specific Interactions Vol. 70, pp. 119–170.
Wassmer, K.-H. see Müller, K.: Vol. 95, pp. 1–56.
Ward, I. M.: Determination of Molecular Orientation by Spectroscopic Techniques. Vol. 66, pp. 81–116.
Ward, I. M.: The Preparation, Structure and Prooperties of Ultra-High Modulus Flexible Polymers. Vol. 70, pp. 1–70.
Weidner, R. see *Nuyken, O.*: Vol. 73/74, pp. 145–200.
Werber, F. X.: Polymerization of Olefins on Supported Catalysts. Vol. 1, pp. 180–191.
Wichterle, O., Šebenda, J., and *Králiček, J.*: The Anionic Polymerization of Caprolactam. Vol. 2, pp. 578–595.
Wilkes, G. L.: The Measurement of Molecular Orientation in Polymeric Solids. Vol. 8, pp. 91–136.
Wilkes, G. L. see Yorkgitis, E. M. Vol. 72, pp. 79–110.
Williams, G.: Molecular Aspects of Multiple Dielectric Relaxation Processes in Solid Polymers. Vol. 33, pp. 59–92.
Williams, J. G.: Applications of Linear Fracture Mechanics. Vol. 27, pp. 67–120.
Wöhrle, D.: Polymere aus Nitrilen. Vol. 10, pp. 35–107.
Wöhrle, D.: Polymer Square Planar Metal Chelates for Science and Industry. Synthesis, Properties and Applications. Vol. 50, pp. 45–134.
Wöhrle, D. see Kaneko, M.: Vol. 84, pp. 141–228.
Wolf, B. A.: Zur Thermodynamik der enthalpisch und der entropisch bedingten Entmischung von Polymerlösungen. Vol. 10, pp. 109–171.
Wolf, B. A. see Ballauff, M.: Vol. 84, pp. 1–31.
Wong, C. P.: Application of Polymer in Encapsulation of Electronic Parts. Vol. 84, pp. 63–84.
Woodward, A. E. and *Sauer, J. A.*: The Dynamik Mechanical Properties of High Polymers at Low Temperatures. Vol. 1, pp. 114–158.
Worsfold, D. J. see Beylen, M. van: Vol. 86, pp. 87–143.
Wunderlich, B.: Crystallization During Polymerization. Vol. 5, pp. 568–619.
Wunderlich, B. and *Baur, H.*: Heat Capacities of Linear High Polymers. Vol. 7, pp. 151–368.
Wunderlich, B. and *Grebowicz, J.*: Thermotropic Mesophases and Mesophase Transitions of Linear, Flecible Macromolecules. Vol. 60/61, pp. 1–60.
Wunderlich, B., Möller, M., Grebowicz, J. and *Baur, H.*: Conformational Motion and Disorder in Low and High Molecular Mass Crystals. Vol. 87, pp. 1–121.
Wrasidlo, W.: Thermal Analysis of Polymers. Vol. 13, pp. 1–99.

Yamashita, Y.: Random and Black Copolymers by Ring-Opening Polymerization. Vol. 28, pp. 1–46.
Yamazaki, N.: Electrolytically Initiated Polymerization. Vol. 6, pp. 377–400.
Yamazaki, N and *Higashi, F.*: New Condensation Polymerizations by Means of Phosphorus Compounds. Vol. 38, pp. 1–25.

Yilgör, I. and *McGrath, J. E.*: Polysiloxane Containing Copolymers: A Survey of Recent Developments. Vol. 86, pp. 1–86.
Yokoyama, Y. and *Hall, H. K.*: Ring-Opening Polymerization of Atom-Bridged and Bond-Bridged Bicyclic Ethers, Acetals and Orthoesters. Vol. 42, pp. 107–138.
Yorkgitis, E. M., *Eiss, N. S. Jr.*, *Tran, C.*, *Wilkes, G. L.* and *McGrath, J. E.*: Siloxane-Modified Epoxy Resins. Vol. 72, pp. 79–110.
Yoshida, H. and *Hayashi, K.*: Initiation Process of Radiation-induced Ionic Polymerization as Studied by Electron Spin Resonance. Vol. 6, pp. 401–420.
Young, R. N., *Quirk, R. P.* and *Fetters, L. J.*: Anionic Polymerization of Non-Polar Monomers Involving Lithium. Vol. 56, pp. 1–90.
Yuki, H. and *Hatada, K.*: Sterospecific Polymerization of Alpha-Substituted Acrylic Acid Esters. Vol. 31, pp. 31, pp. 1–45.

Zachmann, H. G.: Das Kristallisations- und Schmelzverhalten hochpolymerer Stoffe. Vol. 3, pp. 581–687.
Zaikov, G. E. see Aseeva, R. M. Vol. 70, pp. 171–230.
Zakharov, V. A., *Bukatov, G. D.*, and *Yermakov, Y. I.*: On the Mechanism of Olifin Polymerization by Ziegler-Natta Catalysts. Vol. 51, pp. 61–100.
Zambelli, A. and *Tosi, C.*: Stereochemistry of Propylene Polymerization. Vol. 15, pp. 31–60.
Zhirkov, P. V. see Malkin, A. Ya.: Vol. 95, pp. 111–148.
Zucchini, U. and *Cecchin, G.*: Control of Molecular-Weight Distribution in Polyolefins Synthesized with Ziegler-Natta Catalytic Systems. Vol. 51, pp. 101–154.
Zweifel, H. see Lohse, F.: Vol. 78, pp. 59–80.

Subject Index

Adhesion, matrix separation 37
–, separation of fibers 21
– to matrix 20
Adhesive interaction, influence on conductivity 127
Anisodiametrical particles 88, 93

Bubble dynamics 106
– escape, two-phase flow 102, 113
– nuclei 103
–, volume and size 106
Blowing agent, gas-filled polymers 102

Casson equation 74
Composite filler, abrasive activity 50
– –, agglomeration 43
– –, critical concentration 9
– –, interphase conception 11
– –, – thickness 8, 10
– –, partial size 7, 16, 18, 22
– –, structural network 5, 25
– –, surface 38, 42
Composite materials, base polymer 3
– –, clusters 31
– –, filler 3, 12, 27, 34
– –, matrix defects 9, 29, 40
– –, molecular chain orientation 13, 56
– – polymer matrix 5, 14, 17, 24, 36, 50, 55, 58
– –, polymerization filling 44, 46, 48, 49, 51
Conducting composites, applications of 142–145
– –, constituents 127, 128
– –, mechanical properties 141
– – for shielding electromagnetic radiation 144, 145
Conductivity *see* Resistivity

Defects in conducting chains 130, 131
Distribution of fillers in composites 130
– – – –, influence of processing on 132, 134
– – – –, injection-molded samples 134
– – – –, polymer blends 136–138
Dynamic modulus, amplitude dependence 93, 95

Elastic modulus 70, 93
Extrudates, thermoplastic melts 99ff
– structure, temperature gradient 118

Fibers, glass 11, 19, 20, 25, 27, 28
–, –, diameter 11, 28
–, –, fibrous fillers 27, 28
–, –, length 19, 20
–, organic 36
Filled polymer melts, stress states 81
– polymers, rheology 69ff
Fillers, conducting 127–128
– –, aspect ratio 128
– –, classification 128
– –, destruction in processing 138–140
– –, redistribution in processing 133–134
Filler, dispersion medium 90
Fillers, gas 100
Filler, effect on viscosity of filled system 83
– elasticity 92
Fillers, neutral 136
–, –, use in conducting composites 136
Filler particles, anisodiametric 89, 93
– –, nonspherical 88, 93
– – shape 81
– – size, suspension viscosity 89
Fillers, rheological properties 69ff
Filler, shape of particle 81
– surface, viscosity 90
Fillers, uniaxial extension 81
Fluid volumetric flow rates 113
Foaming, chemical 101
Foam plastics 100
–, small-cellular extrudate 101
Free gas phase 102

Gas bubbles, low-molecular liquids, rheology 113
– –, pressure 110
Gas-containing melts, study of flow 110
– polymers, flow structures 117f
– –, rheology 99ff
Gas dissolved in melt, single-phase flow 102, 109
Gas-filled plastics 100ff
Gas-liquid nonpolymer mixtures, flow 117
Glass fibers 11, 19, 20, 25, 27, 28

Injection 55

Local pressure deviations 110

Percolation theory 129, 130
– threshold 129, 130
Polymerization filling 140
– –, conducting composites produced by 140–142
Polymer foams 100
– matrix 127, 129
– –, influence on conduction 127, 129
Porous plastics 100

Quantum mechanical tunneling 129, 130

Rheological equation of state 70
– properties, complex modulus 15
– –, dynamic modulus 9, 45
– –, elasticity modulus 15, 35, 52
– –, modulus 30
– –, shear modulus 5
Resistivity, anisotropy of 136
–, influence of different factors on 128–131
–, – – polymer matrix on 127, 129
–, – – processing conditions on 131, 133
–, – – shear deformations 138
–, mechanism of 128–131
–, profile in sample cross-section 134–137
–, unreproducibility during processing 137

Shear flow, normal stress
– rate 70
– –, wall slip 87
– stress 70
– viscosity 70
Shielding electromagnetic radiation, conducting composites for 143–145
Single-phase flows 109
Statistical systems, distribution of fillers 130
Structurized systems, distribution of fillers 130

Thermoplastic melts, extrusion 99ff
– –, gas-containing 99ff
Two-phase flows 113

Viscosity 4, 11, 23, 32, 48, 56
–, baric factor 19
–, dispersion 82
–, effective 25, 33
–, melt flow index 6
– of dispersion medium 77, 82
–, relative 26

Wall slip, disperse filler 87

Yield stress 70, 71ff
– –, filler concentration 80
– –, temperature 77
– –, viscosity of dispersion medium 77